I0484574

NUREG-1556
Vol. 4

Consolidated Guidance About Materials Licenses

Program-Specific Guidance About Fixed Gauge Licenses

Final Report

Manuscript Completed: October 1998
Date Published: October 1998

Prepared by
P. J. Henderson, A. S. Kirkwood, S. H. Lewis,
W. H. Radcliffe, G. M. Watson

Division of Industrial and Medical Nuclear Safety
Office of Nuclear Material Safety and Safeguards
U.S. Nuclear Regulatory Commission
Washington, DC 20555-0001

ABSTRACT

As part of its redesign of the materials licensing process, the Nuclear Regulatory Commission (NRC) is consolidating and updating numerous guidance documents into a single comprehensive repository as described in NUREG-1539, "Methodology and Findings of the NRC's Materials Licensing Process Redesign," dated April 1996, and draft NUREG-1541, "Process and Design for Consolidating and Updating Materials Licensing Guidance," dated April 1996. NUREG-1556, Vol. 4, "Consolidated Guidance about Materials Licenses: Program-Specific Guidance about Fixed Gauges Licenses," dated October 1998, is the fourth program-specific guidance developed for the new process and is intended for use by applicants, licensees, and NRC staff and will also be available to Agreement States. This document supersedes the guidance found in Draft Regulatory Guide and Value/Impact Statement, FC 404-4, "Guide for the Preparation of Applications for Licenses for the Use of Sealed Sources and Nonportable Gauging Devices," dated January 1985, in NMSS Policy and Guidance Directive (P&GD), FC 85-4, "Standard Review Plan for Applications for Use of Sealed Sources and Nonportable Gauging Devices," dated February 6, 1985, and in NMSS P&GD, FC 85-8, Revision (Rev.) 1, "Licensing of Fixed Gauges and Similar Devices," dated June 29, 1988. This final report takes a more risk-informed, performance-based approach to licensing fixed gauges, and reduces the information (amount and level of detail) needed to support an application to use these devices. It incorporates many suggestions received during the comment period on draft NUREG-1556, Vol. 4. When published, this final report should be used in preparing fixed gauge license applications. NRC staff will use this final report in reviewing these applications.

CONTENTS

APPENDICES

FIGURES

TABLES

FOREWORD

The United States Nuclear Regulatory Commission (NRC) is using Business Process Redesign (BPR) techniques to redesign its materials licensing process. This effort is described in NUREG-1539, "Methodology and Findings of the NRC's Materials Licensing Process Redesign," dated April 1996. A critical element of the new process is consolidating and updating numerous guidance documents into a NUREG-series of reports. Below is a listing of volumes currently included in the NUREG - 1556 series: "Consolidated Guidance About Materials Licenses":

Vol. No.	Volume Title	Status
1	Program-Specific Guidance About Portable Gauge Licenses	Final Report
2	Program-Specific Guidance About Industrial Radiography Licenses	Final Report
3	Applications for Sealed Source and Device Evaluation and Registration	Final Report
4	Program-Specific Guidance About Fixed Gauge Licenses	Final Report
5	Program-Specific Guidance about Self-Shielded Irradiator Licenses	Draft for Comment
6	Program-Specific Guidance about 10 CFR Part 36 Irradiators	Draft for Comment
7	Program-Specific Guidance about Academic, Research and Development, and Other Licenses of Limited Scope	Draft for Comment
8	Program-Specific Guidance about Licenses for Exempt Distribution	Final Report
9	Program-Specific Guidance about Medical Use Licenses	Draft for Comment
10	Program-Specific Guidance about Master Materials Licenses	Draft for Comment
11	Program-Specific Guidance about Licenses of Broad Scope	Draft for Comment

The current document (NUREG-1556, Vol. 4, "Consolidated Guidance about Materials Licenses: Program-Specific Guidance about Fixed Gauges," dated October 1998) is the fourth program-specific guidance developed for the new process. It is intended for use by applicants, licensees, NRC license reviewers, and other NRC personnel. It supersedes the guidance for applicants and licensees previously found in Draft Regulatory Guide and Value/Impact Statement, FC 404-4, "Guide for the Preparation of Applications for Licenses for the Use of Sealed Sources in Nonportable Gauging Devices," dated January 1985, and the guidance for licensing staff previously found in P&GD, FC 85-4, "Standard Review Plan for Applications for the Use of Sealed Sources in Nonportable Gauging Devices," dated February 6, 1985, P&GD, FC 85-8, Rev. 1, "Licensing of Fixed Gauges and Similar Devices," dated June 29, 1988, and the documents marked with an asterisk (*) in Appendix Q. NUREG-1556, Vol. 4 incorporates suggestions submitted during the comment period on draft NUREG-1556, Vol. 4. See the Addendum for summaries of comments, staff responses, and other changes.

FOREWORD

This report takes a risk-informed, performance-based approach to licensing fixed gauges, i.e., it reduces the amount of information needed from an applicant seeking to possess and use a relatively safe device. These fixed gauges containing sealed sources of radioactive material incorporate features engineered to enhance their safety. NRC's considerable experience with these licensees indicates that radiation exposures to workers are generally low, if workers follow basic safety procedures, and the gauges operate as designed.

A team composed of NRC staff from headquarters and regional offices drafted this document, drawing on their collective experience in radiation safety in general and as specifically applied to fixed gauges. A representative of NRC's Office of the General Counsel provided a legal perspective.

NUREG-1556, Vol. 4, "Consolidated Guidance about Materials Licenses: Program-Specific Guidance about Fixed Gauges," dated October 1998, represents a step in the transition from the current paper-based process to the new electronic process. This document is available on the Internet at the following uniform resource locator (URL):
<http://www.nrc.gov/NRC/NUREGS/SR1556/V4/index.html>.

The performance-based approach in NUREG-1556, Vol. 4, gives licensees greater flexibility than previously permitted under licenses based on applications prepared according to Draft Regulatory Guide and Value/Impact Statement, FC 404-4, "Guide for the Preparation of Applications for Licenses for the Use of Sealed Sources in Nonportable Gauging Devices," dated January 1985. This guidance document permits licensees to make more changes in their radiation safety program without amending their licenses, thus reducing the regulatory burden on licensees and the NRC staff. Accordingly, existing fixed gauge licensees have the option of submitting a complete application using NUREG-1556, Vol. 4, at the time that they file an amendment request. Licensees choosing this option should incorporate the requested change into the complete application, submit it with the appropriate amendment fee, and indicate that the complete revision is an amendment request to take advantage of the new guidance. When the staff has reviewed the request and resolved any outstanding issues, the staff will amend the license without changing the expiration date.

Licensees wishing to renew their licenses should submit a complete application according to NUREG-1556, Vol. 4. The staff's action will be similar to that described for amendments, but will include an extension of the license's expiration date. By following this procedure, the staff expects all existing fixed gauge licenses to be converted to the more performance-based format within a few years.

This report describes and makes available to the public information on: methods acceptable to the NRC staff for implementing specific parts of the Commission's regulations; techniques the staff uses in evaluating applications, including specific problems or postulated accidents; and data the NRC staff needs to review applications for licenses. NUREG-1556, Vol. 4, "Consolidated Guidance about Materials Licenses: Program-Specific Guidance about Fixed Gauges," dated October 1998, is not a substitute for NRC regulations, and compliance is not required. The approaches and methods described in this report are provided for information only. Methods and solutions different from those described in this report will be acceptable if they provide a basis for the staff to make the determinations needed to issue or continue a license.

Frederick C. Combs, Acting Director
Division of Industrial and Medical Nuclear Safety
Office of Nuclear Material Safety and Safeguards

ACKNOWLEDGMENTS

The writing team thanks the individuals listed below for assisting in the development and review of both the draft and final reports. All participants provided valuable insights, observations, and recommendations.

The team also thanks Kay Avery, Kathryn M. LaViolette, D. W. Benedict Llewellyn, Alyce J. Martin, Carla T. Purvis, Steven W. Schawaroch, and Gina G. Thompson of Computer Sciences Corporation and Angela S. Case of Total Systems Solutions, Inc.

The Participants

Blough, A. Randolph
Camper, Larry W.
Caniano, Roy J.
Combs, Frederick C.
Cool, Donald A.
Henderson, Pamela J.
Johansen, Jenny M.
Jones, Cynthia G.
Kirkwood, Anthony S.
Lewis, Stephen H.
Merchant, Sally L.
Piccone, Josephine M.
Radcliffe, William R.
Roe, Mary Louise
Schwartz, Maria E.
Treby, Stuart A.
Vacca, Patricia C.
Watson, Gidget M.

ABBREVIATIONS

ALARA	as low as is reasonably achievable
Am-241	americium-241
ANSI	American National Standards Institute
AU	authorized user
bkg	background
BPR	business process redesign
Bq	Becquerel
CaF2	calcium fluoride
Cf-252	californium-252
CDE	committed dose equivalent
CEDE	committed effective dose equivalent
CFR	Code of Federal Regulations
Ci	Curie
C/kg	coulomb per kilogram
Co-60	cobalt-60
cpm	counts per minute
Cs-137	cesium-137
d	day
DOE	United States Department of Energy
DOT	United States Department of Transportation
EDE	effective dose equivalent
FDA	United States Food and Drug Administration
ft	foot
GBq	gigabecquerel
G-M	Geiger-Mueller
GPO	Government Printing Office
hr	hour
IN	Information Notice
IP	Inspection Procedure
kg	kilogram
Kr-85	krypton-85
LiF	lithium fluoride
m	meter
mCi	millicurie
mo	month
MOU	memorandum of understanding
mR	milliroentgen
mrem	millirem
mSv	millisievert
NCRP	National Council on Radiation Protection and Measurements
NIST	National Institute of Standards and Technology
NMSS	Office of Nuclear Material Safety and Safeguards
NRC	United States Nuclear Regulatory Commission

ABBREVIATIONS

NVLAP	National Voluntary Laboratory Accreditation Program
OSP	Office of State Programs
P&GD	Policy and Guidance Directive
Q	Quality Factor
R	Roentgen
Rev.	revision
RG	Regulatory Guide
RQ	reportable quantities
RSO	radiation safety officer
SDE	shallow-dose equivalent
Sr-90	strontium-90
SI	International System of Units (abbreviated SI from the French Le Systeme Internationale d'Unites)
SSD	sealed source and device
std	standard
Sv	Sievert
TAR	technical assistance request
TEDE	total effective dose equivalent
TI	transportation index
TLD	thermoluminescent dosimeters
URL	uniform resource locator
wk	week
yr	year

1   PURPOSE OF REPORT

This report provides guidance to an applicant in preparing a fixed gauge license application as well as NRC criteria for evaluating a fixed gauge license application. It is not intended to address the research and development of fixed gauges or the commercial aspects of manufacturing, distribution, and service of such devices. Within this document, the phrases or terms, "fixed gauge," "gauging devices," or "gauges" are used interchangeably.

This report addresses a variety of radiation safety issues associated with fixed gauges of many designs. Figure 1.1 is a cutaway diagram of a typical fixed gauge showing basic design features. Figure 1.2 illustrates various designs of fixed gauges based, in part, on their intended use and the location of the radioactive source within the gauges. Typically gauges are used for process control (e.g., to measure the thickness of paper, the density of coal, the level of material in vessels and tanks, and volumetric flow rate). Because of differences in design, manufacturers provide appropriate instructions and recommendations for proper operation and maintenance. In addition, with gauges of varying designs, the sealed sources may be oriented in different locations within the devices, resulting in different radiation safety problems.

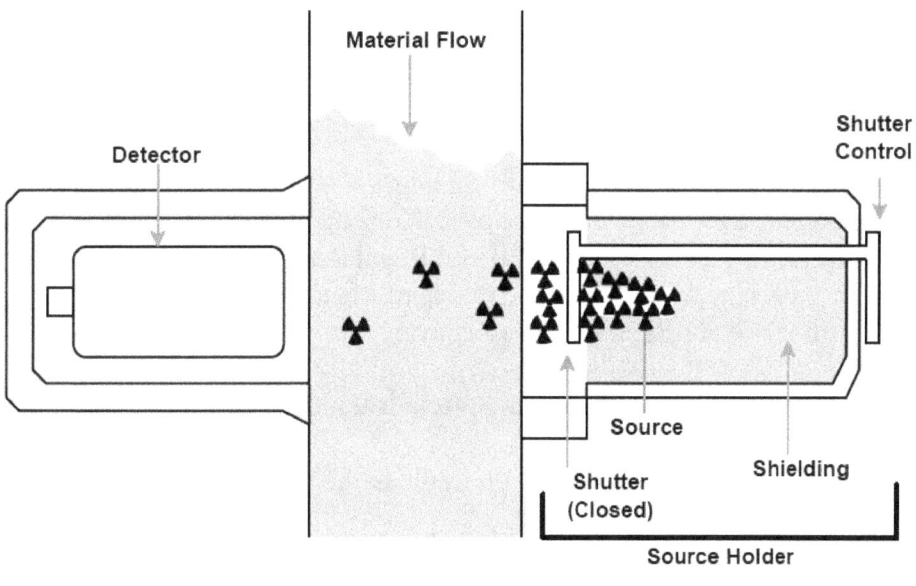

62-pt4-9268-302d
092497

Figure 1.1 Fixed Gauge Basic Design Features. *Cutaway of a typical fixed gauge diagraming the basic design features: the source, source holder, detector, shutter, shutter control or on-off mechanism, and shielding.*

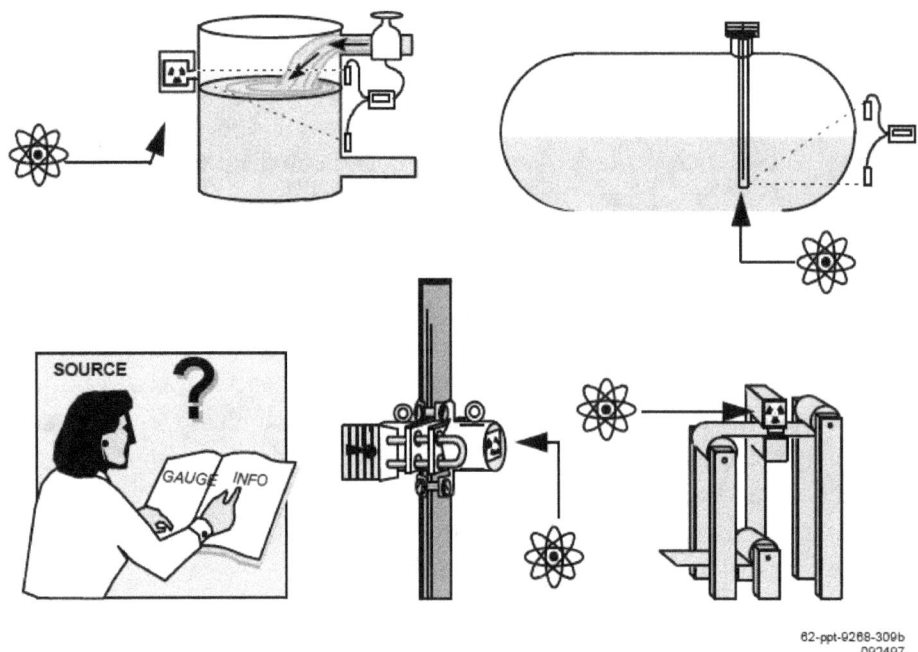

Figure 1.2 Where is the Radioactive Source? *The wide variety of fixed gauge designs results in different radiation safety considerations.*

This report identifies the information needed to complete NRC Form 313 (Appendix A), "Application for Material License," for the use of sealed sources in fixed gauges. The information collection requirements in Title 10, Code of Federal Regulations, Part 30 (10 CFR Part 30) and NRC Form 313 have been approved under the Office of Management and Budget (OMB) Clearance Nos. 3150-0017, and 3150-0120, respectively.

The format within this document for each item of technical information is as follows:

- Regulations — references the regulations applicable to the item;

- Criteria — outlines the criteria used to judge the adequacy of the applicant's response;

- Discussion — provides additional information on the topic sufficient to meet the needs of most readers; and

- Response from Applicant — provides suggested response(s), offers the option of an alternative reply, or indicates that no response is needed on that topic during the licensing process.

Notes and References are self-explanatory and may not be found for each item on NRC Form 313.

Appendix B provides a convenient format to respond to items 5 through 11 on NRC Form 313 (Appendix A) since the form does not have sufficient space for applicants to provide full responses to these items. For streamlined handling of fixed gauge applications in the new materials licensing process, use Appendix B to provide supporting information, attach it to NRC Form 313, and submit them to NRC. Otherwise, as indicated on NRC Form 313, the answers to items 5 through 11 are to be provided on separate sheets of paper and submitted with the completed NRC Form 313.

Appendix C contains information needed for transfer of control. Appendix D is a checklist that NRC staff uses to review applications and applicants can use to check for completeness. Appendix E contains a sample SSD Registration Certificate. Appendixes F through P contain additional information on various radiation safety topics. Appendix Q is a sample fixed gauge license; it contains the conditions most often found on these licenses, although not all licenses will have all conditions. Appendix R contains a list of all documents used to develop this NUREG (NUREG-1556, Vol. 4). The Addendum contains the comments received on draft NUREG 1556, Vol. 4 and NRC staff's response to each comment.

In this document, dose or radiation dose means absorbed dose, dose equivalent, effective dose equivalent (EDE), committed dose equivalent (CDE), committed effective dose equivalent (CEDE), or total effective dose equivalent (TEDE). These terms are defined in 10 CFR Part 20. Rem, and its SI equivalent Sievert (1 rem = 0.01 Sievert (Sv)), are used to describe units of radiation exposure or dose. This is done because 10 CFR Part 20 sets dose limits in terms of rem, not rad or roentgen (R). When the sealed sources used in gauges emit beta and gamma rays, for practical reasons, we are assuming that 1 R = 1 rad = 1 rem. Less common are sealed sources used in gauges that emit neutrons or alpha particles. For neutron and alpha emitting sealed sources, 1 rad is not equal to 1 rem. Determination of dose equivalent (rem) from absorbed dose (rad) from neutrons and alpha particles requires the use of an appropriate quality factor (Q) value. Q values are used to convert absorbed dose (rad) to dose equivalent (rem). Q values for neutrons and alpha particles are addressed in the Tables 1004(b).1 and 2 in 10 CFR §20.1004.

2 AGREEMENT STATES

Certain states, called Agreement States (see Figure 2.1), have entered into agreements with the NRC that give them the authority to license and inspect byproduct, source, or special nuclear materials used or possessed within their borders. A current list of Agreement States (including names, addresses, and telephone numbers of responsible officials) may be obtained upon request from NRC's Regional Offices. Any applicant other than a Federal agency who wishes to possess or use licensed material in one of these Agreement States needs to contact the responsible officials in that State for guidance on preparing an application; file these applications with State officials, not with the NRC.

In general, NRC's materials licensees who wish to conduct operations at temporary job sites in an Agreement State should contact that State's radiation control program office for information about State regulations. To ensure compliance with Agreement State reciprocity requirements, a licensee should request authorization well in advance of scheduled use.

In the special situation of work at Federally-controlled sites in Agreement States, it is necessary to know the jurisdictional status of the land in order to determine whether NRC or the Agreement State has regulatory authority. NRC has regulatory authority over land determined to be "exclusive Federal jurisdiction," while the Agreement State has jurisdiction over non-exclusive Federal jurisdiction land. Licensees are responsible for finding out, in advance, the jurisdictional status of the specific areas where they plan to conduct licensed operations. NRC recommends that licensees ask their local contact for the Federal agency controlling the site (e.g., contract officer, base environmental health officer, district office staff) to help determine the jurisdictional status of the land and to provide the information in writing, so that licensees can comply with NRC or Agreement State regulatory requirements, as appropriate. Additional guidance on determining jurisdictional status is found in All Agreement States Letter, SP-96-022, dated February 16, 1996, which is available as indicated below. Table 2-1 provides a quick way to check on which agency has regulatory authority.

Table 2.1 Who Regulates the Activity?

Applicant and Proposed Location of Work	Regulatory Agency
Federal agency regardless of location (except that Department of Energy [DOE] and, under most circumstances, its prime contractors are exempt from licensing [10 CFR 30.12])	NRC
Non-Federal entity in non-Agreement State, US territory or possession	NRC
AGREEMENT STATES Non-Federal entity in Agreement State at non-Federally controlled site	Agreement State
Non-Federal entity in Agreement State at Federally-controlled site NOT subject to exclusive Federal jurisdiction	Agreement State

Applicant and Proposed Location of Work	Regulatory Agency
Non-Federal entity in Agreement State at Federally-controlled site subject to exclusive Federal jurisdiction	NRC

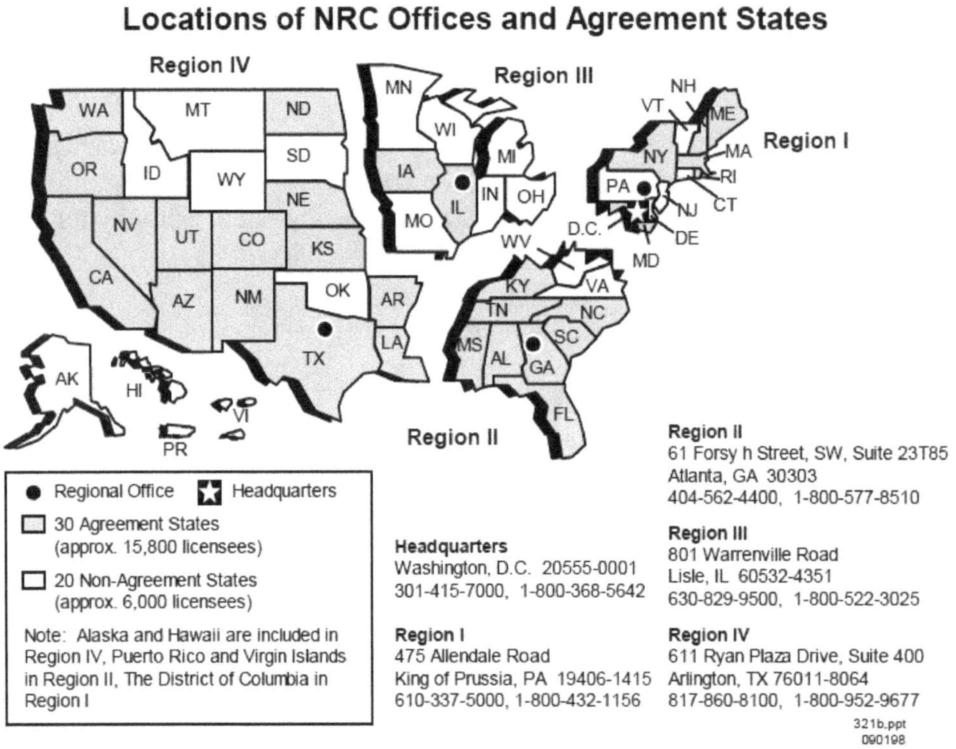

Locations of NRC Offices and Agreement States

Figure 2.1 U.S. Map. Location of NRC Offices and Agreement States.

References: A current list of Agreement States (including names, addresses, and telephone numbers of responsible officials) is available by choosing "Directories" on the NRC Office of State Programs' (OSP's) Home Page <http://www.hsrd.ornl.gov/nrc/home.htm>. As an alternative, request the list from NRC•s Regional Offices.

All Agreement States Letter, SP-96-022, dated February 16, 1996, is available on OSP•s Home Page at <http://www.hsrd.ornl.gov/nrc/home.htm>; choose "NRC-State Communications," then choose "All of the Above" and follow the directions for submitting a query for "SP96022." As an alternative, request the list from OSP; call NRC's toll free number (800) 368-5642 and then ask for extension 415-3340.

3 MANAGEMENT RESPONSIBILITY

The NRC recognizes that effective radiation safety program management is vital to achieving safe and compliant operations. NRC also believes that consistent compliance with its regulations provides reasonable assurance that licensed activities will be conducted safely. NRC also believes that effective management will result in increased safety and compliance.

> "Management" refers to the processes for conduct and control of a radiation safety program and to the individuals who are responsible for those processes and who have *authority to provide necessary resources* to achieve regulatory compliance.

To ensure adequate management involvement, a duly authorized management representative *must* sign the submitted application acknowledging management's commitments and responsibility for the following:

- Radiation safety, security and control of radioactive materials, and compliance with regulations;

- Completeness and accuracy of the radiation safety records and all information provided to NRC (10 CFR 30.9);

- Knowledge about the contents of the license and application;

- Compliance with current NRC and Department of Transportation (DOT) regulations and the licensee's operating and emergency procedures;

- Commitment to provide adequate resources (including space, equipment, personnel, time, and, if needed, contractors) to the radiation protection program to ensure that public and workers are protected from radiation hazards and meticulous compliance with regulations is maintained;

- Selection and assignment of a qualified individual to serve as the Radiation Safety Officer (RSO) for their licensed activities;

- Prohibition against discrimination of employees engaged in protected activities (10 CFR 30.7);

- Commitment to provide information to employees regarding the employee protection and deliberate misconduct provisions in 10 CFR 30.7 and 10 CFR 30.10, respectively;

- Obtaining NRC's prior written consent before transferring control of the license; and

- Notifying appropriate NRC regional administrator in writing, immediately following filing of petition for voluntary or involuntary bankruptcy (10 CFR 30.34(h)).

For information on NRC inspection, investigation, enforcement, and other compliance programs, see the current version of "General Statement of Policy and Procedures for NRC Enforcement Actions," NUREG-1600, and Inspection Procedure (IP) 87110, Appendix A,

"Industrial/Academic/Research Inspection Field Notes." These documents are available electronically at <http://www.nrc.gov>. For hard copies of NUREG-1600 and IP 87110, see the Notice of Availability (on the inside front cover of this report).

4 APPLICABLE REGULATIONS

It is the applicant's or licensee's responsibility to have up-to-date copies of applicable regulations, read them, and abide by each applicable regulation.

The following Parts of 10 CFR Chapter I contain regulations applicable to fixed gauges:

- 10 CFR Part 2, "Rules of Practice for Domestic Licensing Proceedings and Issuance of Orders"

- 10 CFR Part 19, "Notices, Instructions and Reports to Workers: Inspection and Investigations"

- 10 CFR Part 20, "Standards for Protection Against Radiation"

- 10 CFR Part 21, "Reporting of Defects and Noncompliance"

- 10 CFR Part 30, "Rules of General Applicability to Domestic Licensing of Byproduct Material"

- 10 CFR Part 32, "Specific Domestic Licenses to Manufacture or Transfer Certain Items Containing Byproduct Material"

- 10 CFR Part 71, "Packaging and Transportation of Radioactive Material"

Part 71 requires that licensees who transport licensed material or who may offer such material to a carrier for transport must comply with the applicable requirements of the United States Department of Transportation (DOT) that are found in 49 CFR Parts 170 through 189. Copies of DOT regulations can be ordered from the Government Printing Office (GPO) whose address and telephone number are listed below.

- 10 CFR Part 150, "Exemptions and Continued Regulatory Authority in Agreement States and in Offshore Waters Under Section 274"

- 10 CFR Part 170, "Fees for Facilities, Materials, Import and Export Licenses and Other Regulatory Services Under the Atomic Energy Act of 1954, as Amended"

- 10 CFR Part 171, "Annual Fees for Reactor Operating Licenses, and Fuel Cycle Licenses and Materials Licenses, Including Holders of Certificates of Compliance, Registrations, and Quality Assurance Program Approvals and Government Agencies Licensed by NRC"

To request copies of the above documents, call GPO's order desk in Washington, DC at (202) 512-1800. Order the two-volume bound version of Title 10, Code of Federal Regulations, Parts 0-50 and 51-199 from the GPO, Superintendent of Documents, Post Office Box 371954, Pittsburgh, Pennsylvania 15250-7954. You may also contact the GPO electronically at <http://www.gpo.gov>. Request single copies of the above documents from NRC's Regional

Offices (see Figure 2.1 for addresses and telephone numbers). Note that NRC and all other Federal Agencies publish amendments to regulations in the <u>Federal Register</u>.

5 HOW TO FILE

5.1 PAPER APPLICATION

Applicants for a materials license should do the following:

- Be sure to use the most recent guidance in preparing an application.

- Complete NRC Form 313 (Appendix A) Items 1 through 4, 12, and 13 on the form itself.

- Complete NRC Form 313 Items 5 through 11 on supplementary pages or use Appendix B.

- For each separate sheet, other than Appendix B, that is submitted with the application, identify and key it to the item number on the application or the topic to which it refers.

- Submit all documents on 8-1/2 x 11 inch paper.

- Avoid submitting proprietary information unless it is absolutely necessary.

- Submit an original application and one copy.

- Retain one copy of the license application for future reference. Applicants for a materials license should do the following:

> As required by 10 CFR 30.32 (c), applications must be signed by a duly authorized representative; see section in this report entitled "Certification."

> Using the suggested wording of responses and committing to using the model procedures in this report will expedite NRC's review.

All license applications will be available for review by the general public in NRC's Public Document Rooms. If it is necessary to submit proprietary information, follow the procedure in 10 CFR 2.790. Failure to follow this procedure could result in disclosure of the proprietary information to the public or substantial delays in processing the application. Employee personal information, i.e., home address, home telephone number, social security number, date of birth, radiation dose information, should not be submitted unless specifically requested by NRC.

As explained in the Foreword to this document, NRC's new licensing process will be faster and more efficient, in part, through acceptance and processing of electronic applications at some future date. NRC will continue to accept paper applications. However, these will be scanned and put through an optical character reader to convert them to electronic format. To ensure a smooth transition, applicants are requested to follow these suggestions:

- Submit printed or typewritten, not handwritten, text on smooth, crisp paper that will feed easily into the scanner.

- Choose typeface designs that are sans serif, such as Arial, Futura, Univers; the text of this document is in a serif font called **Times New Roman**.

- Choose 12-point or larger font size.

- Avoid stylized characters such as script, italic, etc.

- Be sure the print is clear and sharp.

- Be sure there is high contrast between the ink and paper (black ink on white paper is best).

5.2   ELECTRONIC APPLICATION

As the electronic licensing process develops, it is anticipated that NRC may provide mechanisms for filing applications via diskettes or CD-ROM, and through the Internet. Additional filing instructions will be provided as these new mechanisms become available. The existing paper process will be used until the electronic process is available.

6 WHERE TO FILE

Applicants wishing to possess or use licensed material in any State or U. S. territory or possession subject to NRC jurisdiction must file an application with the NRC Regional Office for the locale in which the material will be possessed and/or used. Figure 2.1 shows NRC's four Regional Offices and their respective areas for licensing purposes and identifies Agreement States.

In general, applicants wishing to possess or use licensed material in Agreements States must file an application with the Agreement State, not NRC. However, if work will be conducted at Federally controlled sites in Agreement States, applicants must first determine the jurisdictional status of the land in order to determine whether NRC or the Agreement State has regulatory authority. See the section on "Agreement States" for additional information.

7 LICENSE FEES

Each application for which a fee is specified, including applications for new licenses and license amendments, must be accompanied by the appropriate fee. Refer to 10 CFR 170.31 to determine the amount of the fee. NRC will not issue the new license prior to fee receipt. Once technical review has begun, no fees will be refunded; application fees will be charged regardless of the NRC's disposition of an application or the withdrawal of an application.

Most NRC licensees are also subject to annual fees; refer to 10 CFR 171.16. Consult 10 CFR 171.11 for additional information on exemptions from annual fees and 10 CFR 171.16(c) on reduced annual fees for licensees that qualify as "small entities."

Direct all questions about NRC's fees or completion of Item 12 of NRC Form 313 (Appendix A) to the Office of the Chief Financial Officer at NRC Headquarters in Rockville, Maryland, (301) 415-7554. You may also call NRC's toll free number (800) 368-5642 and then ask for extension 415-7554.

8 CONTENTS OF AN APPLICATION

The following comments apply to the indicated items on NRC Form 313 (Appendix A).

8.1 ITEM 1: LICENSE ACTION TYPE

THIS IS AN APPLICATION FOR (Check appropriate item):

Type of Action	License No.
[ ] A. New License	Not Applicable
[ ] B. Amendment	XX-XXXXX-XX
[ ] C. Renewal	XX-XXXXX-XX

Check box A for a new license request.

Check box B for an amendment[1] to an existing license, and provide license number.

Check box C for a renewal[1] of an existing license, and provide license number.

8.2 ITEM 2: APPLICANT'S NAME AND MAILING ADDRESS

List the legal name of the applicant's corporation or other legal entity with direct control over use of the radioactive material; a division or department within a legal entity may not be a licensee. An individual may be designated as the applicant only if the individual is acting in a private capacity and the use of the radioactive material is not connected with employment in a corporation or other legal entity. Provide the mailing address where correspondence should be sent. A post office box number is an acceptable mailing address.

Notify NRC of changes in mailing address; these changes do not require a fee.

Note: NRC must be notified before control of the license is transferred or bankruptcy proceedings have been initiated; see below for more details. NRC Information Notice (IN) 97-30, "Control of Licensed Material during Reorganizations, Employee-Management Disagreements, and Financial Crises," dated June 3, 1997, discusses the potential for the security and control of licensed material to be compromised during periods of organizational instability.

---

[1] See "Amendments and Renewals to a License" later in this document. Licensees may request an amendment to an existing license to add authorization for a fixed gauge.

Timely Notification of Transfer of Control

Regulations: 10 CFR 30.34(b).

Criteria: Licensees must provide full information and obtain NRC's **prior written consent** before transferring control of the license, or, as some licensees call it, "transferring the license."

Discussion: Transfer of control may be the result of mergers, buyouts, or majority stock transfers. Although it is not NRC's intent to interfere with the business decisions of licensees, it is necessary for licensees to obtain prior NRC written consent before the transaction is finalized. This is to ensure the following:

- Radioactive materials are possessed, used, or controlled only by persons who have valid NRC licenses;

- Materials are properly handled and secured;

- Persons using these materials are competent and committed to implementing appropriate radiological controls;

- A clear chain of custody is established to identify who is responsible for disposition of records and licensed material; and

- Public health and safety are not compromised by the use of such materials.

Response from Applicant: None from an applicant for a new license; Appendix C, excerpted from IN 89-25, Revision 1, "Unauthorized Transfer of Ownership or Control of Licensed Activities," dated December 7, 1994, identifies the information to be provided about transfers of control.

References: See Notice of Availability (on the inside front cover of this report) to obtain copies of IN 89-25, Revision 1, "Unauthorized Transfer of Ownership or Control of Licensed Activities," dated December 7, 1994, and IN 97-30, "Control of Licensed Material during Reorganizations, Employee-Management Disagreements, and Financial Crises," dated June 3, 1997.

Notification of Bankruptcy Proceedings

Regulations: 10 CFR 30.34(h).

Criteria: Immediately following filing of voluntary or involuntary petition for bankruptcy for or against a licensee, the licensee must notify the appropriate NRC Regional Administrator, in writing, identifying the bankruptcy court in which the petition was filed and the date of filing.

Discussion: Even though a licensee may have filed for bankruptcy, the licensee remains responsible for all regulatory requirements. NRC needs to know when licensees are in bankruptcy proceedings in order to determine whether all licensed material is accounted for and adequately controlled and whether there are any public health and safety concerns (e.g., contaminated facility). NRC shares the results of its determinations with other involved entities (e.g., trustee) so that health and safety issues can be resolved before bankruptcy actions are completed.

Response from Applicant: None at time of application for a new license. Generally, licensees should notify NRC within 24 hours of filing a bankruptcy petition.

References: INs are available in the "Reference Library" on NRC's Home Page at <http://www.nrc.gov>. For hard copies, see the Notice of Availability (on the inside front cover of this report).

8.3 ITEM 3: ADDRESS(ES) WHERE LICENSED MATERIAL WILL BE USED OR POSSESSED

Specify the street address, city, and state or other descriptive address (such as on Highway 10, 5 miles east of the intersection of Highway 10 and State Route 234, Anytown, State) for each facility. The descriptive address should be sufficient to allow an NRC inspector to find the facility location. A Post Office Box address is not acceptable.

> An NRC license does not relieve a licensee from complying with other applicable Federal, State, or local requirements (e.g., local zoning requirements or local ordinances requiring registration of radioactive material).

An NRC-approved license amendment is required before locating a gauge at an address *not* already listed on the license, whether that gauge is an additional unit or a relocation of an existing unit.

For information on conducting operations at temporary job sites (i.e., locations where work is conducted for limited periods of time, refer to the section in this report called "Fixed Gauges Used at Temporary Job Sites." That section offers examples of operations where fixed gauges might be used at temporary job sites and gives information which should be provided to the NRC to support a request for these operations.

> The applicant need *not* submit sketches or identify the specific location of the fixed gauge within the facility with the application. The acceptability of the gauge's location will be reviewed during the inspection process.

Note: As discussed later in the section "Financial Assurance and Record Keeping for Decommissioning," licensees do need to maintain permanent records on where licensed material was used or stored while the license was in force. This is important for making future determinations about the release of these locations for unrestricted use (e.g., before the license is terminated). For fixed gauge licensees, acceptable records are sketches or written descriptions of specific locations where each gauge was used or stored and any information relevant to damaged devices or leaking radioactive sources.

8.4   ITEM 4: PERSON TO BE CONTACTED ABOUT THIS APPLICATION

Identify the individual who can answer questions about the application and include his or her telephone number. This is typically the proposed radiation safety officer, unless the applicant has named a different person as the contact. The NRC will contact this individual if there are questions about the application.

Notify NRC if the contact person or his or her telephone number changes so that NRC can contact the applicant or licensee in the future with questions, concerns, or information. This notice is for "information only" and does not require a license amendment or a fee.

As indicated on NRC Form 313 (Appendix A), Items 5 through 11 should be submitted on separate sheets of paper. Applicants may use Appendix B for this purpose and should note that using the suggested wording of responses and committing to using the model procedures in this report will expedite NRC's review. Appendix D provides examples of the criteria that NRC license reviewers use to review submissions of alternative procedures.

8.5   ITEM 5: RADIOACTIVE MATERIAL

8.5.1   SEALED SOURCES AND DEVICES

Regulations: 10 CFR 30.32(g), 10 CFR 30.33(a)(2), 10 CFR 32.210.

Criteria: Applicants must provide the manufacturer's or distributor's name and model number for each requested sealed source and device. Licensees will be authorized to possess and use only those sealed sources and devices specifically approved or registered by NRC or an Agreement State.

Discussion: NRC or an Agreement State performs a safety evaluation of fixed gauges before authorizing a manufacturer or distributor to distribute the gauges to specific licensees. The safety evaluation is documented in a Sealed Source and Device (SSD) Registration Certificate. Before the SSD registration process was formalized, some older gauges may not have been evaluated in a separate document, but were specifically approved on a license. Licensees can continue to use these gauges that are specifically listed on their licenses. Some examples of fixed gauges are shown in Figure 8.1.

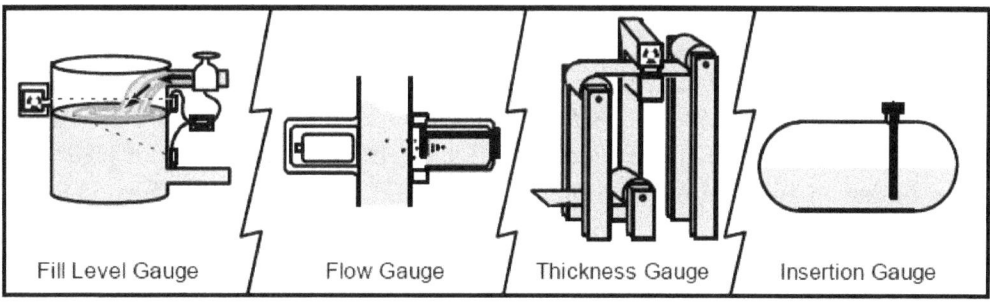

Fill Level Gauge   Flow Gauge   Thickness Gauge   Insertion Gauge

62-ppt-9268-307d
091797

Figure 8.1 Examples of Several Different Types of Fixed Gauges.

Consult with the proposed manufacturer or distributor to ensure that requested sources and devices are compatible and conform to the sealed source and device designations registered with NRC or an Agreement State. Licensees may not make any changes to the sealed source, device, or source/device combination that would alter the description or specifications from those indicated in the respective registration certificates, without obtaining NRC's prior permission in a license amendment. Such changes may necessitate a custom registration review, increasing the time needed to process a licensing action.

SSD Registration Certificates contain sections on "Conditions of Normal Use" and "Limitation and Other Considerations of Use." These sections may include limitations derived from conditions imposed by the manufacturer or distributor, by particular conditions of use that would reduce radiation safety of the device, or by circumstances unique to the sealed source or device. For example, working life of the device or appropriate temperature and other environmental conditions may be specified. Except as specifically approved by NRC, licensees are required to use gauges according to their respective SSD Registration Certificates. Accordingly, applicants may want to obtain a copy of the certificate and review it with the manufacturer or distributor or with NRC or the issuing Agreement State to ensure that it correctly reflects the radiation safety properties of the source or device. See Appendix E for an example of a fixed gauge SSD Registration Certificate.

Response from Applicant:

- Identify each radionuclide that will be used in each source in the gauging device(s).

- Identify the manufacturer or distributor and model number of each type of sealed source and device requested.

- Confirm that each sealed source, device, and source/device combination is registered as an approved sealed source or device by NRC or an Agreement State.

- Confirm that the activity per source and maximum activity per device will not exceed the maximum activity listed on the approved certificate of registration issued by NRC or by an Agreement State.

Note: For information on SSD registration certificates, contact the Registration Assistant by calling NRC's toll free number (800) 368-5642 and then asking for extension 415-7217. For more information about the SSD registration process, see the current version of NUREG-1556, Vol. 3, "Consolidated Guidance About Materials Licenses: Applications for Sealed Source and Device Evaluation and Registration." It is available electronically in the "Reference Library" at <http://www.nrc.gov>; for a paper copy, see the Notice of Availability (on the inside front cover of this report).

8.5.2   FINANCIAL ASSURANCE AND RECORDKEEPING FOR DECOMMISSIONING

Regulations: 10 CFR 30.34(b), 10 CFR 30.35.

Criteria: Fixed gauge licensees authorized to possess sealed sources containing radioactive material in excess of the limits specified in 10 CFR 30.35(b) and (d) must provide evidence of financial assurance for decommissioning.

Even if no financial assurance is required, licensees are required to maintain, in an identified location, decommissioning records related to structures and equipment where gauges are used or stored and to leaking sources. Pursuant to 10 CFR 30.35(g), licensees must transfer these records important to decommissioning to either of the following:

- The new licensee before licensed activities are transferred or assigned according to 10 CFR 30.34(b).

- The appropriate NRC regional office before the license is terminated.

Discussion: The requirements for financial assurance are specific to the types and quantities of byproduct material authorized on a license. Most fixed gauge applicants and licensees do not need to take any action to comply with the financial assurance requirements because their total inventory of licensed material does not exceed the thresholds in 10 CFR 30.35(b) and (d). The thresholds for typical radionuclides used for fixed gauge sealed sources are shown in Table 8.1.

Table 8.1 Examples of Minimum Inventory Quantities Requiring Financial Assurance

Radionuclide (Sealed Sources)	Activity in Gigabecquerels	Activity in Curies
Co-60	3.7×10^5	10,000
Kr-85	3.7×10^7	1,000,000
Sr-90	3.7×10^4	1,000
Cs-137	3.7×10^6	100,000
Am-241	3.7×10^3	100
Cf-252	3.7×10^3	100

A licensee would need to possess hundreds of gauges before the financial assurance requirements would apply. Since the standard gauge license does not specify the maximum number of gauges that a licensee may possess (allowing flexibility in obtaining additional gauges specifically authorized by the license as needed without amending its license), it contains a condition requiring the licensee to limit its possession of fixed gauges to quantities not requiring financial assurance. Applicants and licensees desiring to possess gauges exceeding the threshold amounts must submit evidence of financial assurance.

Applicants requesting more than one radionuclide may determine whether financial assurance for decommissioning is required by calculating, for each radionuclide possessed, the ratio between the activity possessed, in curies, and the radionuclide's threshold activity requiring financial assurance, in curies. If the sum of such ratios for all of the radionuclides possessed exceeds "1" (i.e., "unity"), then applicants must submit evidence of financial assurance for decommissioning.

The same regulation also requires that licensees maintain records important to decommissioning in an identified location. All fixed gauge licensees need to maintain records of structures and equipment where each gauge was used or stored. As-built drawings with modifications of structures and equipment shown as appropriate fulfill this requirement. If drawings are not available, licensees shall substitute appropriate records (e.g., a sketch of the room or building or a narrative description of the area) concerning the specific areas and locations. If no records exist regarding structures and equipment where gauges were used or stored, licensees shall make all reasonable efforts to create such records based upon historical information (e.g. employee recollections). In addition, if fixed gauge licensees have experienced unusual occurrences (e.g., leaking sources, other incidents that involve spread of contamination), they also need to maintain records about contamination that remains after cleanup or that may have spread to inaccessible areas.

> For fixed gauge licensees whose sources have never leaked, acceptable records important to decommissioning are sketches or written descriptions of the specific locations where each gauge was used or stored.

Response from Applicant: No response is needed from most applicants. If financial assurance is required, submit the documentation required under 10 CFR 30.35. RG 3.66, "Standard Format and Content of Financial Assurance Mechanisms Required for Decommissioning Under 10 CFR Parts 30, 40, 70, and 72," dated June 1990, contains approved wording for each of the mechanisms authorized by the regulation to guarantee or secure funds except for the Statement of Intent for Government licensees.

> Licensees must transfer records important to decommissioning either to the new licensee before licensed activities are transferred or assigned in accordance with 10 CFR 30.34(b) or to the appropriate NRC regional office before the license is terminated.

References: See Notice of Availability (on the inside front cover of this report) to obtain copies of RG 3.66, "Standard Format and Content of Financial Assurance Mechanisms Required for Decommissioning Under 10 CFR Parts 30, 40, 70, and 72," dated June 1990, and P&GD FC 90-2, Revision 1, "Standard Review Plan for Evaluating Compliance with Decommissioning Requirements," dated April 30, 1991.

8.6 ITEM 6: PURPOSE(S) FOR WHICH LICENSED MATERIAL WILL BE USED

Regulations: 10 CFR 30.33(a)(1).

Criteria: Proposed purpose is authorized by the Atomic Energy Act of 1954, as amended. Gauges should be used only for the purposes for which they were designed, according to the manufacturer's or distributor's recommendations and instructions, as specified in an approved SSD Registration Certificate, and as authorized on an NRC license.

Discussion: Uses other than those listed in the SSD Registration Certificate require review and approval by the NRC or an Agreement State. Requests to use fixed gauges for purposes not listed in the SSD Registration Certificate will be reviewed on a case-by-case basis. Applicants need to submit sufficient information to demonstrate that the proposed use will not compromise the integrity of the source or source shielding, or other radiation safety-critical components of the device. NRC will evaluate the radiation safety program for each type and use of gauge requested.

> An NRC license does not relieve a licensee from complying with other applicable Federal, State, or local regulations.

Response from Applicant: Provide either of the following:

- If the fixed gauge(s) will be used for the purposes listed on the SSD Registration Certificate, do the following:

  — State that "The fixed gauge(s) will be used for the purposes described on the SSD Registration Certificate(s)"

  — Provide a specific description of use for each type of gauge requested, e.g., "for use in measuring the thickness of paper, the bulk density and weight of coal on a beltscale, etc."

OR

- If the fixed gauge will be used for purposes other than those listed on the SSD Registration Certificate, specify these other purposes and submit safety analyses (and procedures, if needed) to support safe use.

Note:

- Allowed uses of fixed gauges normally include process control methods such as measuring the thickness of paper, the density of coal, the level of material in vessels and tanks, etc.

- Unusual uses will be evaluated on a case-by-case basis and the authorized use condition will reflect approved uses.

8.7 ITEM 7: INDIVIDUAL(S) RESPONSIBLE FOR RADIATION SAFETY PROGRAM AND THEIR TRAINING EXPERIENCE

8.7.1 RADIATION SAFETY OFFICER (RSO)

Regulations: 10 CFR 30.33(a)(3).

Criteria: Radiation Safety Officers (RSOs) must have adequate training and experience. Successful completion of training of one of the following is evidence of adequate training and experience.

- Fixed gauge manufacturer's or distributor•s course for users or for RSOs

- Equivalent course that meets Appendix G criteria

Additional training is required for RSOs of programs that perform non-routine operations. This includes repairs involving or potentially affecting components related to the radiological safety of the gauge (e.g., the source, source holder, source drive mechanism, shutter, shutter control, or shielding) and any other activities during which personnel could receive radiation doses exceeding NRC limits (e.g., installation, initial radiation survey, gauge relocation, and removal of the gauge from service). See "Radiation Safety Program - Maintenance" in this report and Appendix N, "Non Routine Operations."

Discussion: The person responsible for the radiation protection program is called the Radiation Safety Officer (RSO). The RSO needs independent authority to stop operations that he or she considers unsafe. He or she must have sufficient time and commitment from management to fulfill certain duties and responsibilities to ensure that radioactive materials are used in a safe manner. Typical RSO duties are illustrated in Figure 8.2 and described in Appendix F. NRC requires the name of the RSO on the license to ensure that licensee management has always identified a responsible, qualified person and that the named individual knows of his or her designation as RSO.

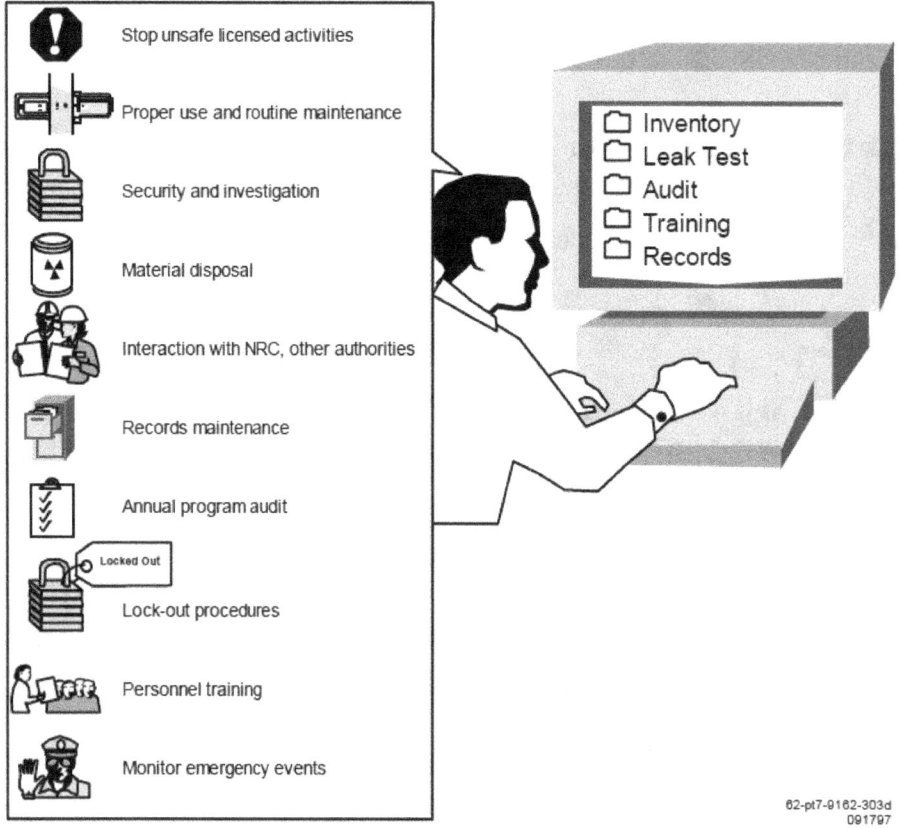

62-pt7-9162-303d
091797

Figure 8.2 RSO Responsibilities. *Typical duties and responsibilities of RSOs.*

Response from Applicant: Provide the following:

- Name of the proposed RSO;

AND EITHER

- Statement that: "Before obtaining licensed materials, the proposed RSO will have successfully completed the training described in Criteria in the section entitled 'Radiation Safety Officer' in NUREG-1556, Vol. 4, 'Consolidated Guidance about Materials Licenses: Program-Specific Guidance about Fixed Gauges Licenses,' dated October 1998"; and

- Statement that: "Before being named as the RSO, future RSOs will have successfully completed the training described in Criteria in the section entitled 'Radiation Safety Officer' in NUREG-1556, Vol. 4, 'Consolidated Guidance about Materials Licenses: Program-Specific Guidance about Fixed Gauges Licenses,' dated October 1998. Within 30 days of naming a new RSO, we will submit the new RSO's name to NRC to include in our license."

OR

- Alternative information demonstrating that the proposed RSO and any future RSO are qualified by training and experience.

Note:

- It is important to notify NRC, as soon as possible, of changes in the designation of the RSO; such notifications will be handled as administrative amendments not requiring fees as long as the application contains the commitment listed in the third bullet under "Response from Applicant."

- Alternative responses will be evaluated using the criteria listed above.

8.7.2   AUTHORIZED USERS

Regulations: 10 CFR 30.33(a)(3).

Criteria: Authorized users (AUs) must have adequate training and experience. Successful completion of one of the following is evidence of adequate training and experience:

- Fixed gauge manufacturer's or distributor•s course for users

- Equivalent course that meets Appendix G criteria

Applicants requesting to perform non-routine operations such as installation, initial radiation survey, repair, and maintenance of components related to the radiological safety of the gauge,

gauge relocation, replacement and disposal of sealed sources, alignment, or removal of a gauge from service, must provide additional training. See the section in this report entitled "Radiation Safety Program - Maintenance" and Appendix N.

Discussion: An AU is a person whose training and experience meet NRC criteria, who is named either explicitly or implicitly on the license, and who uses or directly supervises the use of licensed material. AUs must ensure the proper use, security, and routine maintenance of fixed gauges containing licensed material. AUs must attend the training and instruction given at the time of installation or receive equivalent training and instruction.

An AU is considered to be supervising the use of licensed material when he or she directs personnel in operations involving the material. Although the AU may delegate specific tasks to supervised users (e.g., maintaining records), he or she is still responsible for safe use of licensed material.

Response from Applicant: Provide either of the following:

- The statement: "Before using licensed materials, authorized users will have successfully completed one of the training courses described in Criteria in the section entitled 'Authorized Users' in NUREG-1556, Vol. 4, 'Consolidated Guidance about Materials Licenses: Program-Specific Guidance about Fixed Gauge Licenses,' dated October 1998."

OR

- A description of the training and experience for proposed authorized users.

Note: Alternative responses will be evaluated using the criteria listed above.

8.8 ITEM 8: TRAINING FOR INDIVIDUALS WHO IN THE COURSE OF EMPLOYMENT ARE LIKELY TO RECEIVE OCCUPATIONAL DOSES OF RADIATION IN EXCESS OF 1 mSv (100 mrem) IN A YEAR (OCCUPATIONALLY EXPOSED WORKERS) AND ANCILLARY PERSONNEL

Regulations: 10 CFR 19.11, 10 CFR 19.12, 10 CFR 19.13, 10 CFR 20.1801, 10 CFR 20.1802, 10 CFR 30.7, 10 CFR 30.9, 10 CFR 30.10, 10 CFR 30.33.

Criteria: Individuals who in the course of employment are likely to receive occupational doses of radiation in excess of 1 mSv (100 mrem) in a year must receive training according to 10 CFR 19.12. The extent of this training must be commensurate with potential radiological health protection problems present in the work place.

Discussion: Licensees need to perform a prospective evaluation to determine radiation doses likely to be received by different individuals or groups. AUs, and individuals performing installations, relocations, non-routine maintenance, or repairs would be most likely to receive doses in excess of 1 mSv (100 mrem) in a year. See the previous section for a discussion of training and experience for AUs.

Licensee personnel who work in the vicinity of a fixed gauge but do not use gauges (ancillary staff) are not required to have radiation safety training as long as they are not likely to receive 1 mSv (100 mrem) in a year. However, to minimize potential radiation exposure when ancillary staff are working in the vicinity of a fixed gauge, it is prudent for them to work under the supervision and in the physical presence of an AU or to be provided some basic radiation safety training. Such ancillary staff should be informed of the nature and location of the gauge and the meaning of the radiation symbol, and should be instructed not to touch the gauge and to keep away from it as much as their work permits.

Some ancillary staff, although not likely to receive doses over 1 mSv (100 mrem), should receive training to ensure adequate security and control of licensed material. Licensees may provide these individuals with training commensurate with their assignments in the vicinity of the gauge, to ensure the control and security of licensed material.

Response from Applicant: The applicant's training program, for individuals who in the course of employment are likely to receive occupational doses of radiation in excess of 1 mSv (100 mrem) in a year (occupationally exposed workers) and ancillary personnel, will be examined during inspections, but should not be submitted in the license application.

8.9   ITEM 9: FACILITIES AND EQUIPMENT

Regulations: 10 CFR 30.33(a)(2), 10 CFR 32.210.

Criteria: Facilities and equipment must be adequate to protect health and to minimize danger to life or property. This may be demonstrated by the following:

- The location of the gauge is compatible with the "Conditions of Normal Use" and "Limitations and/or Other Considerations of Use" on the SSD Registration Certificate

- The fixed gauge is secured to prevent unauthorized removal or access (e.g., located in a locked room, permanently mounted, or chained and locked to a storage rack).

Discussion: Fixed gauges incorporate many engineering features to protect the user from unnecessary radiation exposure in a wide variety of environments. Fixed gauges may be located in harsh environments involving variables such as pressure, vibration, mounting height/method, temperature, humidity, air quality, corrosive atmospheres, corrosive chemicals including process materials and cleaning agents, possible impact or puncture conditions, and fire, explosion, and flooding potentials. Applicants need to consult the sections on the SSD Registration Certificate entitled, "Conditions of Normal Use" and "Limitations and/or Other Considerations of Use" to determine the appropriate gauge for a location. In those instances when a proposed location is not consistent with the SSD Registration Certificate, the applicant may ask the source or device manufacturer or distributor to request an amendment to modify the SSD Registration Certificate to include the new conditions. If the manufacturer or distributor does not request an amendment, the applicant must provide the NRC with specific information demonstrating that the proposed new conditions will not impact the safety or integrity of the source or device.

Response from Applicant: Provide one of the following:

- A statement that: "We will ensure that the location of each fixed gauge meets the criteria in the section entitled 'Facilities and Equipment' in NUREG-1556, Vol. 4, 'Consolidated Guidance about Materials Licenses: Program-Specific Guidance about Fixed Gauge Licenses,' dated October 1998."

OR

- Confirm that the fixed gauge is secured to prevent unauthorized removal or access; and submit specific information demonstrating that the proposed conditions will not impact the safety or integrity of the source or device. Address any instances where the proposed conditions exceed any conditions listed in the SSD Registration Certificate.

Note:

- Any deviations from an SSD Registration Certificate will require specific NRC approval.

- Alternative responses will be evaluated using the criteria listed above.

References: INs are available in the "Reference Library" on NRC's Home Page at <http://www.nrc.gov>. For hard copies, see the Notice of Availability (on the inside front cover of this report).

8.10   ITEM 10: RADIATION SAFETY PROGRAM

8.10.1   AUDIT PROGRAM

Regulations: 10 CFR 20.1101, 10 CFR 20.2102.

Criteria: Licensees must review the content and implementation of their radiation protection programs at intervals not to exceed 12 months to ensure the following:

*   Compliance with NRC and DOT regulations (as applicable), and the terms and conditions of the license;

*   Occupational doses and doses to members of the public are ALARA (10 CFR 20.1101); and

*   Records of audits and other reviews of program content are maintained for 3 years.

Discussion: Appendix H contains a suggested audit program that is specific to the use of fixed gauges and is acceptable to NRC. All areas indicated in Appendix H may not be applicable to every licensee and all items may not need to be addressed during each audit. For example, licensees do not need to address areas which do not apply to their activities, and activities which have not occurred since the last audit need not be reviewed at the next audit.

Currently the NRC's emphasis in inspections is to perform actual observations of work in progress. As a part of their audit programs, applicants should consider performing unannounced audits of fixed gauge users to determine if, for example, Operating and Emergency Procedures are available and are being followed, etc.

It is essential that once identified, problems be corrected comprehensively and in a timely manner; IN 96-28, "Suggested Guidance Relating to Development and Implementation of Corrective Action," provides guidance on this subject. The NRC will review the licensee's audit results and determine if corrective actions are thorough, timely, and sufficient to prevent recurrence. If violations are identified by the licensee and these steps are taken, the NRC will normally exercise discretion and may elect not to cite a violation. The NRC's goal is to encourage prompt identification and prompt, comprehensive correction of violations and deficiencies. For additional information on NRC's use of discretion on issuing violations, refer to the current version of NUREG-1600, "General Statement of Policy and Procedures for NRC Enforcement Actions."

Licensees must maintain records of audits and other reviews of program content and implementation for 3 years from the date of the record. NRC has found audit records that contain the following information to be acceptable: date of audit, name of person(s) who conducted audit, persons contacted by the auditor(s), areas audited, audit findings, corrective actions, and follow-up.

Response from Applicant: The applicant's program for reviewing the content and implementation of its radiation protection program will be examined during inspections, and should not be submitted in the license application.

References: The current version of NUREG-1600 is available electronically at <http://www.nrc.gov/OE>. INs are available in the "Reference Library" on NRC's Home Page at <http://www.nrc.gov>. For hard copies of NUREG-1600, IN 96-28, and IP 87110, Appendix A, "Industrial/Academic/Research Inspection Field Notes," see the Notice of Availability (on the inside front cover of this report).

8.10.2 INSTRUMENTS

Regulations: 10 CFR 20.1501, 10 CFR 20.2103(a), 10 CFR 30.33(a)(2).

Criteria: Licensees must possess, or have access to, radiation monitoring instruments which are necessary to protect health and minimize danger to life or property. Instruments used for quantitative radiation measurements must be calibrated periodically for the radiation measured.

Discussion: Usually it is not necessary for fixed gauge licensees to possess a survey meter. However, surveys according to 10 CFR 20.1501 will be required if an applicant plans to conduct non-routine operations. This includes installation, initial radiation surveys, relocation, removal from service, dismantling, alignment, replacement, disposal of the sealed source, and non-routine maintenance and repair of components related to the radiological safety of the gauge. Because some of these operations may increase the risk of radiation exposure, individuals performing these operations should be carefully monitored with a survey meter. Such survey meters should be properly calibrated. Proper calibration is particularly important for initial surveys since the results can be used as a basis for public dose estimates. For those licensees requesting authorization to calibrate their own survey instruments, Appendix I contains calibration procedures acceptable to the NRC.

Licensees who perform surveys pursuant to 10 CFR 20.1501 must possess a survey meter that:

- Measures at least 0.3 through 1 through 200 mR per hour (50 microcoulombs per kilogram)

- Is capable of measuring the radiation being emitted from the gauge•s sealed source

- Is checked for functionality with a source of radiation at the beginning of each day of use (e.g., with the gauge or a check source)

- Is calibrated:

  — At intervals not to exceed 12 months

  — Using a source of radiation similar to those found in the gauges

— After any servicing or repair (other than a simple battery exchange)

— To ensure that exposure rates indicated by the meter do not vary from the actual exposure rates by more than ± 20% on each scale

— By the instrument manufacturer or person specifically authorized by the NRC or an Agreement State

Since many fixed gauge licensees are not required to possess a survey meter, applicants should preplan how they will obtain assistance in performing a radiation survey in the event of an emergency (e.g., obtain a survey instrument from hospitals, universities, other NRC or Agreement State licensees, or local emergency response organization). It is important to determine as soon as possible after an incident, by the use of a radiation survey meter, whether the shielding and source are intact.

For those licensees using gauges containing only beta, neutron or alpha-emitting radionuclides, specialized survey instruments may be required.

Response from Applicant: Provide one of the following:

- A statement that, "Surveys according to 10 CFR 20.1501 will be performed by a person specifically authorized by the NRC or an Agreement State to perform these surveys."

OR

- A statement that, "We will use survey instruments that meet the Criteria in the section entitled 'Radiation Safety Program - Instruments' in NUREG-1556, Vol. 4, 'Consolidated Guidance about Materials Licenses: Program-Specific Guidance about Fixed Gauge Licenses,' dated October 1998"; and one of the following *three* choices:

  — "Each survey meter will be calibrated by the manufacturer or other person authorized by the NRC or an Agreement State to perform survey meter calibrations."

OR

  — "We will implement the model survey meter calibration program published in Appendix I entitled 'Survey Instrument Calibration' in NUREG-1556, Vol. 4, 'Consolidated Guidance about Materials Licenses: Program-Specific Guidance about Fixed Gauge Licenses,' dated October 1998"

OR

  — "We will submit alternative calibration procedures for NRC review."

OR, IN LIEU OF ALL OF THE ABOVE

- Submit a description of an alternative method to perform surveys according to 10 CFR 20.1501.

Notes:

- Alternative responses will be reviewed against the criteria listed above.

- The NRC license will state that survey meter calibrations will be performed by the instrument manufacturer or a person specifically authorized by the NRC or an Agreement State, unless the applicant specifically requests this authorization. Applicants seeking authorization to perform survey meter calibrations must submit additional information for review. See Appendix I for more information.

- Regardless of whether an applicant is authorized to calibrate survey meters or contracts an authorized firm to perform calibrations, the licensee must retain calibration records for at least 3 years.

8.10.3 MATERIAL RECEIPT AND ACCOUNTABILITY

Regulations: 10 CFR 30.34(e), 10 CFR 30.41, 10 CFR 30.51, 10 CFR 20.1801, 10 CFR 20.1802, 10 CFR 20.2201.

Criteria: Licensees must do the following:

- Maintain records of receipt, transfer, and disposal of fixed gauges and

- Conduct physical inventories at intervals not to exceed 6 months, or some other interval justified by the applicant and approved by NRC, to account for all sealed sources.

Discussion: As illustrated in Figure 8.3, licensed materials must be tracked from "cradle to grave" in order to ensure gauge accountability, identify when gauges could be lost, stolen, or misplaced, and ensure that possession limits listed on the license are not exceeded. Significant problems can arise from failure to ensure the accountability of gauges. See IN 88-02, "Lost or Stolen Gauges," dated February 2, 1988.

Cradle to Grave Accountability

62-pt4-9162-301c
110397

Figure 8.3 Material Receipt and Accountability. *Licensees must maintain records of receipt, transfer, and disposal and conduct semiannual physical inventories.*

Receipt, transfer, and disposal records must be maintained for the times specified in Table 8-2. Typically, these records contain the following types of information:

- Radionuclide and activity (in units of becquerels or curies) of byproduct material in each sealed source

- Manufacturer's or distributor's name, model number, and serial number (if appropriate) of each device containing byproduct material

- Location of each sealed source and device

- For materials transferred or disposed of, the date of the transfer or disposal, name and license number of the recipient, description of the affected radioactive material (e.g., radionuclide, activity, manufacturer's or distributor's name and model number, serial number).

Table 8.2   Record Maintenance

Type of Record	How Long Record Must be Maintained
Receipt	For as long as the material is possessed until 3 years after transfer or disposal
Transfer	For 3 years after transfer
Disposal	Until NRC terminates the license
Important to Decommissioning*	Until the site is released for unrestricted use

* See the section entitled "Financial Assurance and Recordkeeping for Decommissioning."

Response from Applicant: Provide either of the following:

- A statement that: "Physical inventories will be conducted at least every 6 months or at other intervals approved by the NRC, to account for all sealed sources and devices received and possessed under the license."

OR

- A description of the procedures for ensuring that no fixed gauge has been lost, stolen, or misplaced and how often they will be conducted.

Note:

- Alternative responses will be evaluated using the criteria listed above.
- Inventory records should be maintained and contain the following types of information:
    — Radionuclide and amount (in units of becquerels or curies) of byproduct material in each sealed source
    — Manufacturer's or distributor•s name, model number, and serial number (if appropriate) of each device containing byproduct material
    — Location of each sealed source and device
    — Date of the inventory
    — Signature of the individual conducting the inventory.

References: INs are available in the "Reference Library" on NRC's Home Page at <http://www.nrc.gov>. For hard copies, see the Notice of Availability (on the inside front cover of this report).

8.10.4   OCCUPATIONAL DOSE

Regulations: 10 CFR 20.1502, 10 CFR 20.1201, 10 CFR 20.1207, 10 CFR 20.1208.

Criteria: Applicants must do either of the following:

- Perform a prospective evaluation demonstrating that unmonitored individuals are not likely to receive, in one year, a radiation dose in excess of 10% of the allowable limits as shown in Figure 8.4

OR

- Provide dosimetry processed and evaluated by a National Voluntary Laboratory Accreditation Program (NVLAP) approved processor that is exchanged at a frequency recommended by the processor.

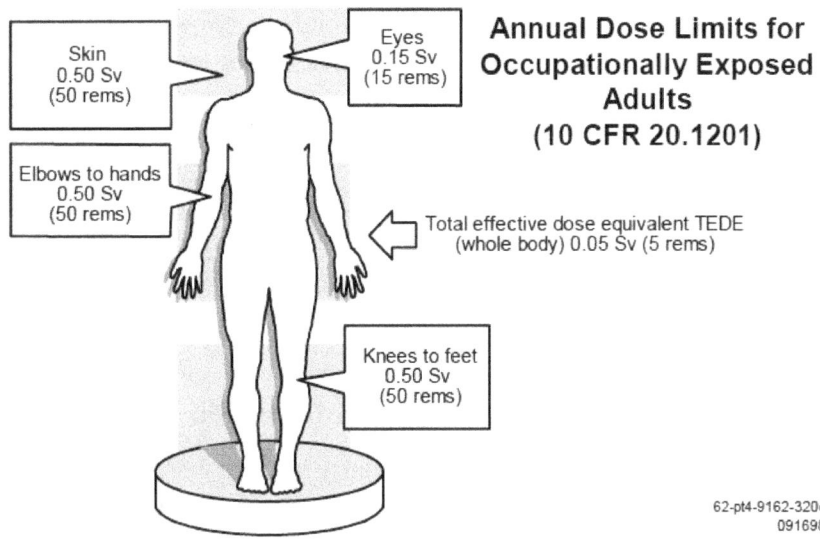

Figure 8.4 Annual Dose Limits for Occupationally Exposed Adults.

Discussion: Under conditions of routine use, the typical fixed gauge user does not require a personnel monitoring device (dosimetry). A gauge user also does not require dosimetry when proper emergency procedures are used. Appendix J provides guidance on performing a prospective evaluation demonstrating that fixed gauge users are not likely to exceed 10% of the limits as shown in Figure 8.4 and thus, are not required to have personnel dosimetry.

Individuals who perform non-routine operations such as installation, initial radiation survey, repair, and maintenance of components related to the radiological safety of the gauge, gauge relocation, replacement, and disposal of sealed sources, alignment, or removal of a gauge from service are more likely to exceed 10% of the limits as shown in Figure 8.4. Applicants may be required to provide dosimetry (whole body and perhaps extremity monitors) to individuals performing such services or must perform a prospective evaluation demonstrating that unmonitored individuals performing such non-routine operations are not likely to receive, in one year, a radiation dose in excess of 10% of the allowable limits as shown in Figure 8.4.

When personnel monitoring is needed, most licensees use either film badges or thermoluminescent dosimeters (TLDs) that are supplied by a NVLAP-approved processor. The exchange frequency for film badges is usually monthly due to technical concerns about film fading. The exchange frequency for TLDs is usually quarterly. Applicants should verify that the processor is NVLAP-approved. Consult the NVLAP-approved processor for its recommendations for exchange frequency and proper use.

Response from Applicant: Provide one of the following:

- A statement that: "We will perform a prospective evaluation demonstrating that unmonitored individuals are not likely to receive, in one year, a radiation dose in excess of 10% of the allowable limits in 10 CFR Part 20 or we will provide dosimetry that meets the Criteria in the section entitled 'Radiation Safety Program - Occupational Dosimetry' in NUREG-1556, Vol. 4, 'Consolidated Guidance about Materials Licenses: Program-Specific Guidance about Fixed Gauge Licenses,' dated October 1998"

OR

- A description of an alternative method for demonstrating compliance with the referenced regulations.

Notes:

- Alternative responses will be evaluated using the criteria listed above.

- Some licensees choose to provide personnel dosimetry to their workers for reasons other than compliance with NRC requirements (e.g., to respond to worker requests).

References: National Institute of Standards and Technology (NIST) Publication 810, "National Voluntary Laboratory Accreditation Program Directory," is published annually and is available electronically at <http://ts.nist.gov/nvlap>. NIST Publication 810 can be purchased from GPO, whose URL is <http://www.gpo.gov>. ANSI N322 may be ordered electronically at <http://www.ansi.org> or by writing to ANSI, 1430 Broadway, New York, NY 10018.

8.10.5 PUBLIC DOSE

Regulations: 10 CFR 20.1301, 10 CFR 20.1302, 10 CFR 20.1003, 10 CFR 20.1801, 10 CFR 20.1802, 10 CFR 20.2107.

Criteria: Licensees must do the following:

- Ensure that fixed gauges will be used, transported, and stored in such a way that members of the public will not receive more than 1 mSv [100 mrem] in one year, and the dose in any unrestricted area will not exceed 0.02 mSv [2 mrem] in any one hour, from licensed operations.

- Prevent unauthorized access, removal, or use of fixed gauges.

Discussion: Public dose is defined in 10 CFR Part 20 as "the dose received by a member of the public from exposure to radiation and/or radioactive material released by a licensee, or to any other source of radiation under the control of a licensee." Public dose excludes doses received from background radiation and from medical procedures. Whether the dose to an individual is an occupational dose or a public dose depends on the individual's assigned duties. It does not depend on the area (restricted, controlled, or unrestricted) the individual is in when the dose is received.

In the case of fixed gauges, members of the public include persons who live, work, or may be near locations where fixed gauges are used or stored and employees whose assigned duties do not include the use of licensed materials and who work in the vicinity where gauges are used or stored. Since a fixed gauge presents a radiation field, the applicant must use methods to limit the public dose such that the radiation level in an unrestricted area (e.g., a nearby walkway or area near the gauge that requires frequent maintenance) does not exceed 1 mSv (100 mrem) in a year or 0.02 mSv (2 mrem) in any one hour.

Because fixed gauges are generally permanently mounted (e.g., chained and locked to a storage rack), they may not need to be in a locked area to prevent loss, theft, or unauthorized removal. Operating and emergency procedures regarding security and lock-out procedures specified in this document should be sufficient to limit the exposure to the public during use or storage and after accidents. IN 81-37, "Unnecessary Radiation Exposures to the Public and Workers During Events Involving Thickness and Level Measuring Devices," dated December 15, 1981, provides information about two events that resulted or may have resulted in unnecessary radiation exposure to members of the public and to maintenance workers. IN 88-02, "Lost or Stolen Gauges," dated February 2, 1988, provides information about several events where fixed gauges were lost or stolen.

Public dose is also affected by the location of the gauge. Use the concepts of time, distance, and shielding when developing a method to limit public dose. Decreasing the time spent near a gauge, increasing the distance from the gauge, and using shielding will reduce the radiation exposure. The most effective way to limit public dose is to prevent members of the public from entering areas where gauges are used or stored. This may be accomplished by administrative or engineering controls.

Administrative controls include training and warning signs. In cases where gauges are located in hostile environments (e.g., high temperatures, caustic chemicals, etc.), warning signs may be difficult to maintain so mandatory training programs may be necessary to caution employees.

Engineering controls reduce radiation levels in areas that are accessible to the public. Shielding the gauge with a protective barrier (e.g., using brick, concrete, lead, or other solid walls) or placing the gauge within an enclosure to prevent access to higher radiation levels are examples of engineering controls. See Figure 8.5.

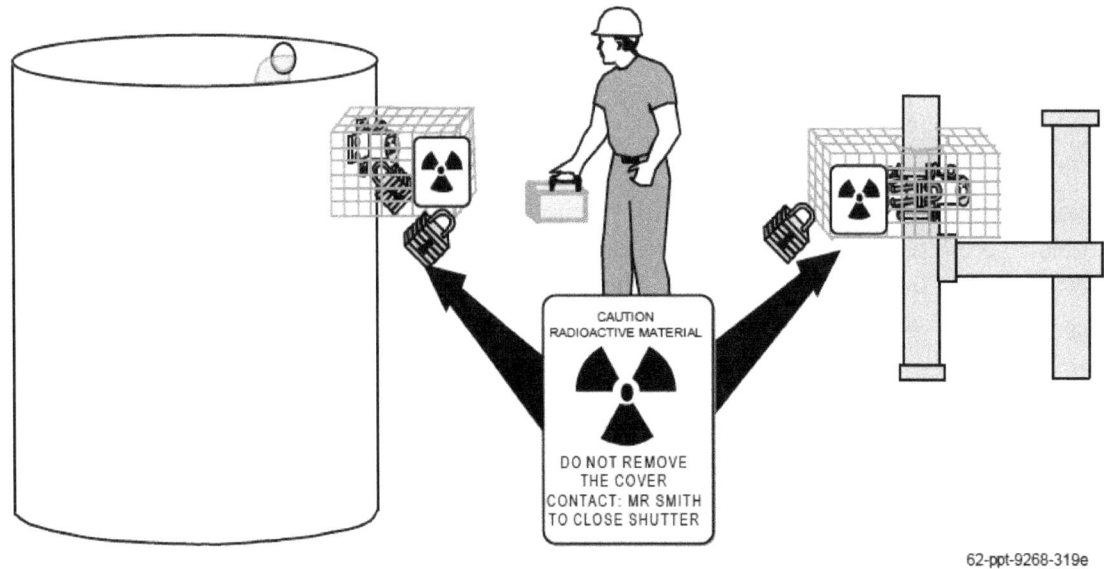

62-ppt-9268-319e
102197

Figure 8.5 Limiting Public Dose. *When dose rates in an area are high enough that a member of the public could receive a dose in excess of 0.02 mSv (2 mrem) in any one hour or 1 mSv (100 mrem) in a year, licensees must take additional measures to prevent public access to these higher dose rates, such as building enclosures around the gauges.*

Public dose can be estimated in areas near the gauge by using radiation levels determined during initial surveys and applying the "inverse square" law to evaluate the effect of distance on radiation levels and occupancy factors to account for the actual presence of members of the public. See Appendix K for an example.

If, after making a public dose estimate, the conditions used to make the evaluation change (e.g., changes the location of gauges, changes the type or frequency of gauge use, adds gauges, changes the occupancy of adjacent areas), then the licensee must perform a new evaluation to ensure that the public dose limits are not exceeded and take corrective action, as needed.

During NRC inspections, licensees must be able to provide documentation demonstrating, by measurement or calculation, that the TEDE to the individual likely to receive the highest dose from the licensed operation does not exceed the annual limit for individual members of the public. See Appendix K for examples of methods to demonstrate compliance.

Response from Applicant: No response is required from the applicant during the licensing phase. Documentation demonstrating compliance will be examined during inspection.

References: See the Notice of Availability (on the inside front cover of this report) to obtain copies of IN 81-37, "Unnecessary Radiation Exposures to the Public and Workers During Events Involving Thickness and Level Measuring Devices," dated December 15, 1981, and IN 88-02, "Lost or Stolen Gauges," dated February 2, 1988.

8.10.6 OPERATING AND EMERGENCY PROCEDURES

Regulations: 10 CFR 30.34(e), 10 CFR 20.1101, 10 CFR 20.1801, 10 CFR 20.1802, 10 CFR 20.2201-2203, 10 CFR 30.50, 10 CFR 21.21, 10 CFR 19.11(a)(3).

Criteria: Each applicant should do the following:

- Develop, implement, and maintain operating procedures containing the following elements for each type of fixed gauge:

    — Instructions for operating the gauge

    — Instructions for performing routine cleaning and maintenance (e.g., calibration and lubrication) according to the manufacturer's or distributor•s recommendations and instructions

    — Instructions for testing each gauge for the proper operation of the on-off mechanism (shutter) and indicator, if any, at intervals not to exceed 6 months or as specified in the SSD certificate

    — Instructions for lock-out procedures, if applicable, that are adequate to assure that no individual or portion of an individual's body can enter the radiation beam

    — Instructions to prevent unauthorized access, removal, or use of fixed gauges

    — Steps to take to keep radiation exposures ALARA

    — Steps to maintain accountability (i.e., inventory)

    — Instructions to ensure that non-routine operations such as installation, initial radiation survey, repair and maintenance of components related to the radiological safety of the gauge, gauge relocation, replacement and disposal of sealed sources, alignment, or removal of a gauge from service are performed by the manufacturer, distributor or person specifically authorized by the NRC or an Agreement State

    — Steps to ensure that radiation warning signs are visible and legible.

- Develop, implement, and maintain emergency procedures for gauge malfunction or damage containing the following elements for each type of fixed gauge:

    — Stop use of the gauge.

    — Restrict access to the area.

    — Contact responsible individuals. (Telephone numbers for the RSO, AUs, the gauge manufacturer or distributor, fire department or other emergency response organization, as appropriate, and the NRC should be posted or easily accessible.)

— Do not attempt repair or authorize others to attempt repair of the gauge except as specifically authorized in a license issued by the NRC or an Agreement State.

— Require timely reporting to NRC pursuant to 10 CFR 20.2201-20.2203, 10 CFR 30.50, and 10 CFR 21.21.

— Take additional steps, dependent on the specific situations.

• Provide copies of operating and emergency procedures to all gauge users.

• Post copies of operating and emergency procedures at each location of use or if posting procedures is not practicable, post a notice which briefly describes the procedures and states where they may be examined.

Discussion: NRC will permit an applicant greater flexibility when licensing certain types of gauges. For each gauge that is requested, if one or more of the following safety conditions are met, the applicant must develop, implement, maintain, and distribute operating and emergency procedures but need *not* submit these procedures for NRC review:

• The air gap between the radiation source and detector of the device is less than 45 cm (18 inches)

• The air gap of the device would not allow insertion of a 30 cm (12 inches) diameter sphere into the radiation beam of the device without removal of a barrier

• The radiation dose rate in the radiation beam of the device at 45 cm (18 inches) from the radiation source with the device shutters, if any, in the open position does not exceed 1 mSv/hour (0.1 rem/hour)

• Entry into vessels (e.g., bins, tanks, hoppers, or pipes) with a gauge installed is not necessary under any foreseeable circumstances and is prohibited.

Operating and emergency procedures should be developed, maintained, and implemented to ensure that gauges are used only as they were designed to be used, control and accountability are maintained, and radiation doses received by occupational workers and members of the public are ALARA. Copies of operating and emergency procedures should be provided to all gauge users. In addition, licensees must post current copies of operating and emergency procedures applicable to licensed activities at each site. If posting of procedures is not practicable, the licensee may post a notice which describes the documents and states where they may be examined.

Improper operation could lead to the damage or malfunction of a gauge and elevated exposure rates in the gauge's immediate vicinity. A list of specific items that should be addressed in operating and emergency procedures is contained in Appendix L. Figure 8.6 illustrates proper response to fire involving a fixed gauge. Emergency procedures should be developed to address a spectrum of incidents (e.g., fire, explosion, mechanical damage, flood, or earthquake).

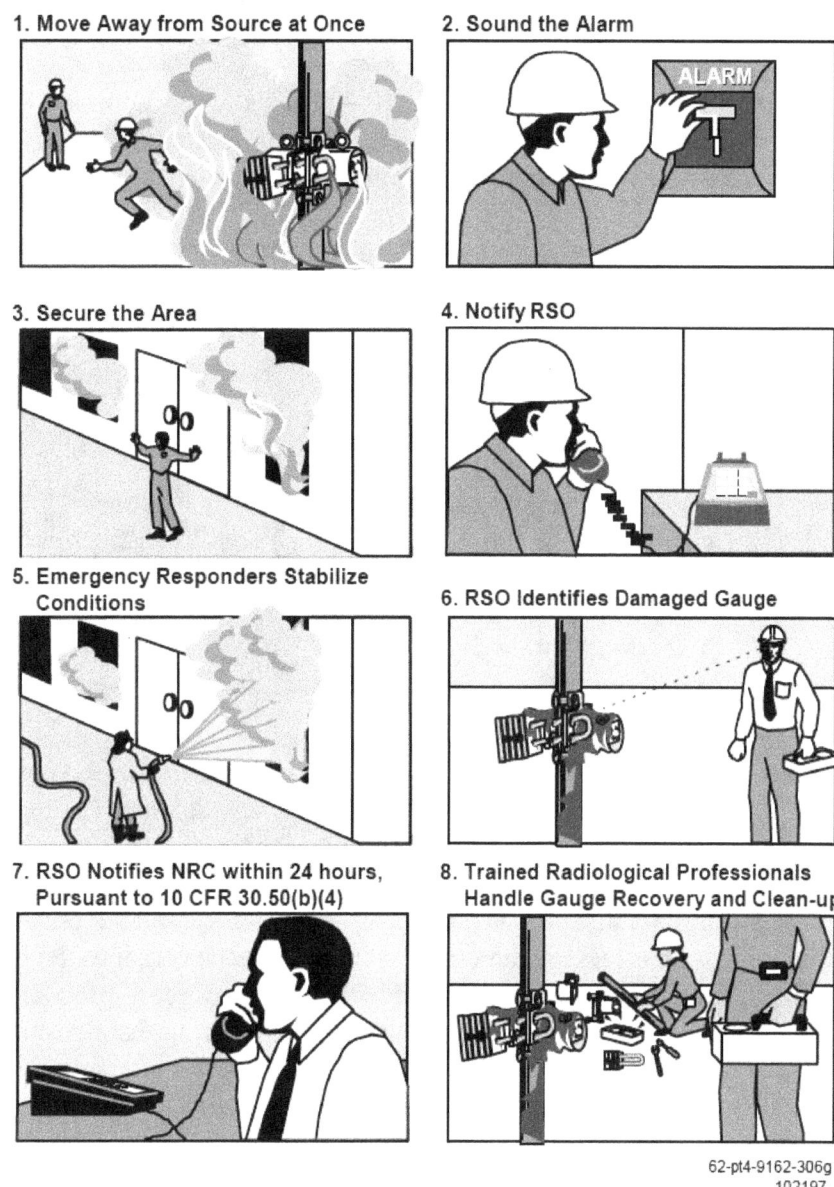

1. Move Away from Source at Once
2. Sound the Alarm
3. Secure the Area
4. Notify RSO
5. Emergency Responders Stabilize Conditions
6. RSO Identifies Damaged Gauge
7. RSO Notifies NRC within 24 hours, Pursuant to 10 CFR 30.50(b)(4)
8. Trained Radiological Professionals Handle Gauge Recovery and Clean-up

62-pt4-9162-306g
102197

Figure 8.6 Proper Handling of Incident. *Licensee personnel implement emergency procedures when a fire melts the lead shielding of a gauge producing the potential for elevated exposure levels.*

NRC considers security of licensed material extremely important and lack of security is a significant violation for which licensees may be fined. Although most fixed gauges are difficult to move, the licensee must prevent unauthorized access, removal, or use of the gauge. Licensees are responsible for ensuring that gauges are secure and accounted for at all times (e.g., during plant modifications, change in ownership, staffing changes, or after termination of activities at a particular location).

The NRC must be notified when gauges are lost, stolen, or certain other conditions occur.

The RSO must be proactive in evaluating whether NRC notification is required. Refer to Appendix P and the regulations (10 CFR 20.2201-20.2203, 10 CFR 30.50, 10 CFR 21.21) for a description of when and where notifications are required.

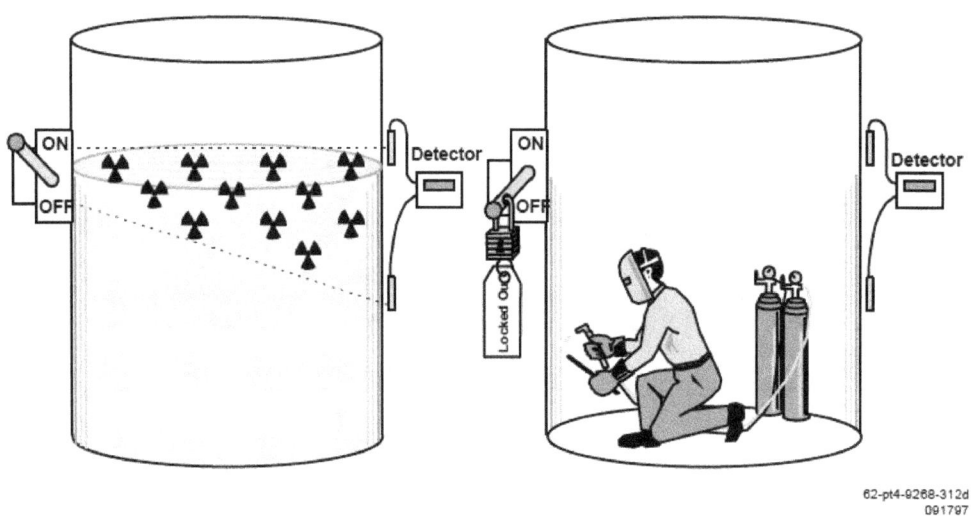

Figure 8.7 Lock-out Procedures. *Typical lock-out procedures include locking the shutter into the "off" position and tagging the shutter control mechanism to indicate the gauge is locked-out.*

When the distance or air gap between the source and detector permits entry of all or a portion of a person's body into the primary radiation beam, licensees must develop lock out procedures. Lock-out procedures encompass locking the on-off or shutter mechanism into the off position or otherwise controlling the radiation beam or using any other means of preventing an individual or a portion of an individual's body from entering the radiation beam during maintenance, repairs, or work in, on, or around the process line (e.g., bin, tank, hopper, pipe, or conveyor belt) where the device is mounted. The on-off or shutter control mechanism should be tagged to indicate that the gauge is locked out. A warning sign should be posted at each entryway to an area where it is possible to be exposed to the primary beam. In addition to providing a warning, the sign should give safety instructions, e.g., "contact the RSO before entering this vessel." Lock-out procedures should specify who is responsible for performing them.

Response from Applicant: Provide either of the following:

- If the gauge meets one or more of the safety conditions specified in "Discussion," provide either of the following:

    — A statement that: "Operating and emergency procedures will be developed, implemented, maintained, and distributed, and will meet the Criteria in the section entitled 'Radiation Safety Program - Operating and Emergency Procedures' in NUREG - 1556, Vol. 4,

'Consolidated Guidance about Materials Licenses: Program-Specific Guidance about Fixed Gauge Licenses,' dated October 1998";

OR

— Submit a description of alternative procedures.

• If the gauge does not meet one or more of the safety conditions specified in "Discussion," provide your operating, emergency, and lock-out (if applicable) procedures to NRC for review.

Note:

• Alternative procedures will be evaluated using the criteria listed above.

• Incidents involving fixed gauges are described in IN 81-37, "Unnecessary Radiation Exposures to the Public and Workers During Events Involving Thickness and Level Measuring Devices," dated December 15, 1981; IN 86-31, "Unauthorized Transfer and Loss of Control of Industrial Nuclear Gauges," dated May 5, 1986; IN 88-02, "Lost or Stolen Gauges," dated February, 2, 1988; IN 88-90, "Unauthorized Removal of Industrial Nuclear Gauges," dated November 22, 1988; and IN 94-15, "Radiation Exposures during an Event Involving a Fixed Nuclear Gauge," dated March 2, 1994. Applicants should consider the information contained in these documents when developing operating and emergency procedures.

References: See the Notice of Availability (on the inside front cover of this report) to obtain copies of IN 81-37, "Unnecessary Radiation Exposures to the Public and Workers During Events Involving Thickness and Level Measuring Devices," dated December 15, 1981; IN 86-31, "Unauthorized Transfer and Loss of Control of Industrial Nuclear Gauges," dated May 5, 1986; IN 88-02, "Lost or Stolen Gauges," dated February 2, 1988; IN 88-90 "Unauthorized Removal of Industrial Nuclear Gauges," dated November 22, 1988; and IN 94-15, "Radiation Exposures during an Event Involving a Fixed Nuclear Gauge," dated March 2, 1994.

8.10.7 LEAK TESTS

Regulations: 10 CFR 30.53, 10 CFR 20.1501, 10 CFR 20.2103.

Criteria: NRC requires testing to determine whether there is any radioactive leakage from the source in the fixed gauge. Records of the test results must be maintained.

Discussion: When issued, a license will require performance of leak tests at intervals approved by the NRC or an Agreement State and specified in the SSD Registration Certificate. The measurement of the leak test sample is a quantitative analysis requiring that instrumentation used to analyze the sample be capable of detecting 185 Bq (0.005 microcurie) of radioactivity.

Manufacturers, distributors, consultants, and other organizations may be authorized by NRC or an Agreement State to either perform the entire leak test sequence for other licensees or provide leak test kits to licensees. In the latter case, the licensee is expected to take the leak test sample according to the fixed gauge manufacturer's and the kit supplier's instructions and return it to the kit supplier for evaluation and reporting results. Leak test samples should be collected at the most accessible area where contamination would accumulate if the sealed source were leaking. See Figure 8.8 below. Licensees may also be authorized to conduct the entire leak test sequence themselves. Appendix M contains information to support a request to perform leak testing and sample analysis.

62-pt4-9268-311f
092497

Figure 8.8 Leak Test Sample. *A leak test sample is collected according to the gauge manufacturer's and the leak test kit supplier's instructions.*

Response from Applicant: Provide one of the following three alternatives:

- A statement that: "Leak tests will be performed at intervals approved by the NRC or an Agreement State and specified in the Sealed Source and Device Registration Certificate. Leak tests will be performed by an organization authorized by NRC or an Agreement State to provide leak testing services to other licensees or using a leak test kit supplied by an organization authorized by NRC or an Agreement State to provide leak test kits to other licensees and according to the kit supplier's instructions. Records of leak test results will be maintained."

OR

- A statement that: "We will implement the model leak test program published in Appendix M to NUREG-1556, Vol. 4, 'Consolidated Guidance about Materials Licenses: Program-Specific Guidance about Fixed Gauge Licenses,' dated October 1998."

OR

- A description of alternative equipment and/or procedures for determining whether there is any radioactive leakage from sources contained in gauges.

Note: Requests for authorization to perform leak testing and sample analysis will be reviewed and, if approved, NRC staff will authorize via a license condition.

References: See the Notice of Availability (on the inside front cover of this report) to obtain copies of Draft Regulatory Guide FC 412-4, "Guide for the Preparation of Applications for the Use of Radioactive Materials in Leak-Testing Services," dated June 1985.

8.10.8 MAINTENANCE

Regulations: 10 CFR 20.1101, 10 CFR 30.34(e).

Criteria: Licensees must routinely clean and maintain gauges according to the manufacturer's or distributor's written recommendations and instructions. Individuals performing routine maintenance must have adequate training and experience. Radiation safety procedures for routine cleaning and maintenance (e.g., removal of exterior residues from the gauge housing, external lubrication of shutter mechanism, calibration, and electronic repairs) must consider ALARA and ensure that the gauge functions as designed and source integrity is not compromised.

Non-routine maintenance or repair (beyond routine cleaning, lubrication, calibration, and electronic repairs) means any maintenance or repair that involves or potentially affects components, including electronics, related to the radiological safety of the gauge (e.g., the source, source holder, source drive mechanism, shutter, shutter control or shielding) and any other activities during which personnel could receive radiation doses exceeding NRC limits.

Non-routine repair or maintenance must be performed by the fixed gauge manufacturer or distributor or a person specifically authorized by NRC or an Agreement State. Information to support requests for specific authorization to perform non-routine maintenance or repair is addressed in Appendix N.

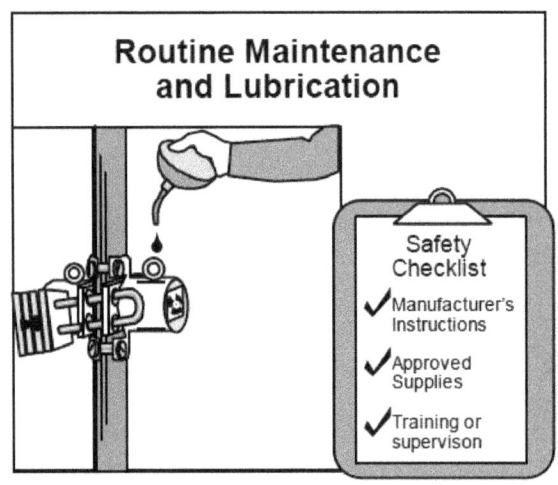
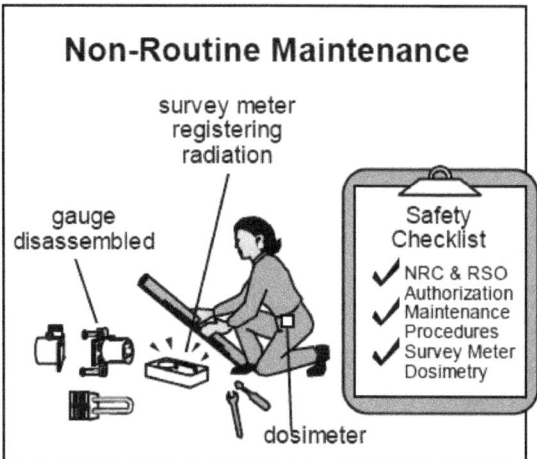

Figure 8.9 Maintenance. *Licensees need to perform routine maintenance to ensure proper operation of the fixed gauge. For non-routine maintenance, most licensees rely on the gauge manufacturer, distributor or other service companies.*

Discussion: NRC permits fixed gauge licensees to perform routine maintenance of the gauges provided that they follow the gauge manufacturer's or distributor•s written recommendations and instructions. Generally, before any maintenance or repair work is done, licensees need to determine (and assure themselves of the adequacy of) the following:

- The tasks to be performed

- The protocol or procedures to be followed

- The radiation safety procedures including possible need for compensatory measures (e.g., steps taken to compensate for lack of or reduced shielding)

- ALARA considerations

- Training and experience of personnel performing the work

- The qualification of parts, components, other materials to be used in the gauge

- The tests (to be performed before the gauge is returned to routine use) to ensure that it functions as designed.

Although manufacturers or distributors may use different terms, "routine maintenance" includes, but is not limited to, cleaning, lubrication, calibration, and electronic repairs.

Routine maintenance does *not* include any activities that involve:

- Components, including electronics, related to the radiological safety of the gauge (e.g., the source, source holder, source drive mechanism, shutter, shutter control or shielding)

- Installation, relocation, or alignment of the gauge

- Initial radiation surveys

- Replacement and disposal of sealed sources

- Removal of a gauge from service

- A potential for any portion of the body to come into contact with the primary radiation beam

- Any other activity during which personnel could receive radiation doses exceeding NRC limits

Mounting a gauge is unpacking or uncrating the gauge, and fastening, hanging, or affixing the gauge into position before using. Mounting does not include electrical connection, activation, or operation of the gauge. Installing a gauge includes mounting, electrical connection, activation, and first use of the device. Specific NRC or Agreement State authorization is required to install a gauge. However, a licensee may initially mount a gauge, without specific NRC or Agreement State authorization, if the gauge's SSD Certificate explicitly permits it and under the following guidelines:

- The gauge must be mounted according to written instructions provided by the manufacturer or distributor

- The gauge must be mounted in a location compatible with the "Conditions of Normal Use" and "Limitations and/or Other Considerations of Use" in the certificate of registration issued by NRC or an Agreement State

- The on-off mechanism (shutter) must be locked in the off position, if applicable, or the source must be otherwise fully shielded

- The gauge must be received in good condition (package was not damaged)

- The gauge must not require any modification to fit in the proposed location.

The source must remain fully shielded and the gauge may not be used until it is installed and made operational by a person specifically licensed by the Commission or an Agreement State to perform such operations.

A condition in the NRC license will state that operations such as installation, initial radiation survey, repair, and maintenance of components related to the radiological safety of the gauge, gauge relocation, replacement, and disposal of sealed sources, alignment, or removal of a gauge from service shall be performed only by the manufacturer, distributor or other persons specifically licensed by the Commission or an Agreement State to perform such services. Most licensees do not perform non-routine operations. Rather, these licensees rely upon persons specifically licensed by the Commission or an Agreement State who have the specialized equipment and technical expertise needed to perform these activities. Applicants seeking authorization to perform non-routine operations must submit specific procedures for review. See Appendix N for more information.

Response from Applicant:

Routine maintenance: Submit either of the following:

- A statement that: "We will implement and maintain procedures for routine maintenance of our gauges according to each manufacturer's or distributor•s written recommendations and instructions."

OR

- Alternative procedures for NRC's review.

Non-routine operations: Submit either of the following:

- A statement that: "The gauge manufacturer, distributor or other person authorized by NRC or an Agreement State will perform non-routine operations such as installation, initial radiation survey, repair, and maintenance of components related to the radiological safety of the gauge, gauge relocation, replacement, and disposal of sealed sources, alignment, or removal of a gauge from service."

OR

- The information listed in Appendix N supporting a request to perform this work "in-house."

Note:

- Alternative procedures for performing routine maintenance will be evaluated using the criteria listed above.
- Information requested in Appendix N will be reviewed on a case-by-case basis; if approved, the license will contain a condition authorizing the licensee to perform non-routine operations.

8.10.9 TRANSPORTATION

Regulations: 10 CFR 71.5, 49 CFR Parts 171-178, 10 CFR 20.1101.

Criteria: Applicants must either:

- Arrange for transportation of a gauge by the manufacturer, distributor or other person specifically licensed to transport gauges by the NRC or Agreement State.

OR

- Develop, implement, and maintain safety procedures for off-site transport of radioactive material to ensure compliance with DOT regulations.

Discussion: Some fixed gauge licensees have the manufacturer, distributor or other person specifically licensed to transport gauges by the NRC or Agreement State arrange for preparing and shipping licensed material. If licensees decide to transport their own gauges, they are responsible for compliance with DOT regulations which require, in part, specific labeling and surveying of the package before shipping. To appropriately survey the package the surveyor must use instruments that can measure radiation exposure rates around the package and detect contamination on the package. Appendix O lists major DOT regulations and provides an example of a shipping paper. During an inspection, NRC uses the provisions of 10 CFR 71.5 and a Memorandum of Understanding (MOU) with DOT on the Transportation of Radioactive Material (signed June 6, 1979) to examine and enforce transportation requirements applicable to gauge licensees.

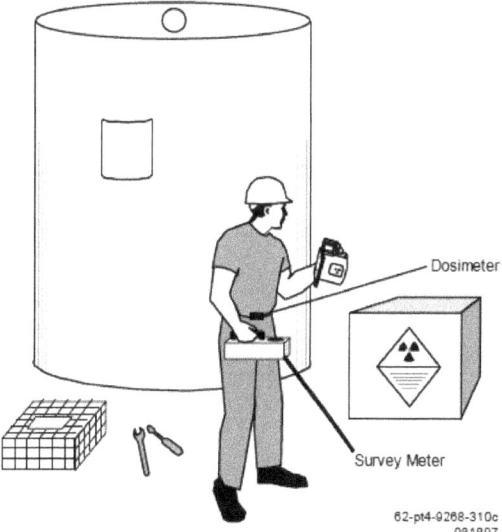

62-pt4-9268-310c
081897

Figure 8.10 Transportation. *Illustration of a fixed gauge being disassembled and packaged for transport.*

Response from Applicant: No response is needed from applicants during the licensing process; this issue will be reviewed during inspection.

References: "A Review of Department of Transportation Regulations for Transportation of Radioactive Materials (1983 revision)" can be obtained by calling DOT's Office of Hazardous Material Initiatives and Training at (202) 366-4425. See the Notice of Availability (on the inside front cover of this report) to obtain copies of the MOU with DOT on the Transportation of Radioactive Material, signed June 6, 1979.

8.10.10 FIXED GAUGES USED AT TEMPORARY JOB SITE

Regulations: 10 CFR 30.34(e), 10 CFR 20.1101, 10 CFR 20.1801, 10 CFR 20.1802, 10 CFR 20.2201-2203, 10 CFR 30.50.

Criteria: Each applicant requesting authorization to perform work with fixed gauges at temporary job sites should do the following:

Develop, implement, maintain, and distribute operating and emergency procedures containing the following elements:

- Instructions for transporting radioactive material to ensure compliance with DOT regulations

- Instructions for using gauges at temporary job sites and performing routine maintenance according to the manufacturer's or distributor•s recommendations and instructions

- Instructions for maintaining security during storage and transportation

- Instructions to keep gauges under control and immediate surveillance or secured to prevent unauthorized use or access.

- Steps to take to keep radiation exposures ALARA

- Steps to maintain accountability during use

- Steps to control access to a potentially damaged gauge (See Figure 8.11)

- Steps to take, and who to contact, when a gauge has been lost or damaged (e.g., local officials, RSO, etc.) (See Figure 8.11)

- If gauges are to be installed at temporary job sites, the operating and emergency procedures should contain instructions on using personal dosimetry and survey instruments and conducting surveys

- Provide copies of operating and emergency procedures to all gauge users and at each job site.

Figure 8.11 Proper Handling of Incident. *Licensee personnel implement emergency procedures when a traffic accident results in a damaged gauge and potentially elevated exposure levels.*

Discussion: A temporary job site is a location where work with licensed materials is conducted for a limited period of time. Temporary job sites are not specifically listed on a license. A gauge user may be dispatched to work at several temporary job sites in one day. A location is not considered a temporary job site if it is used to store *and* dispatch licensed material. The NRC considers such a location to be a field office. Licensees must apply for and receive a license amendment specifically listing each field office location.

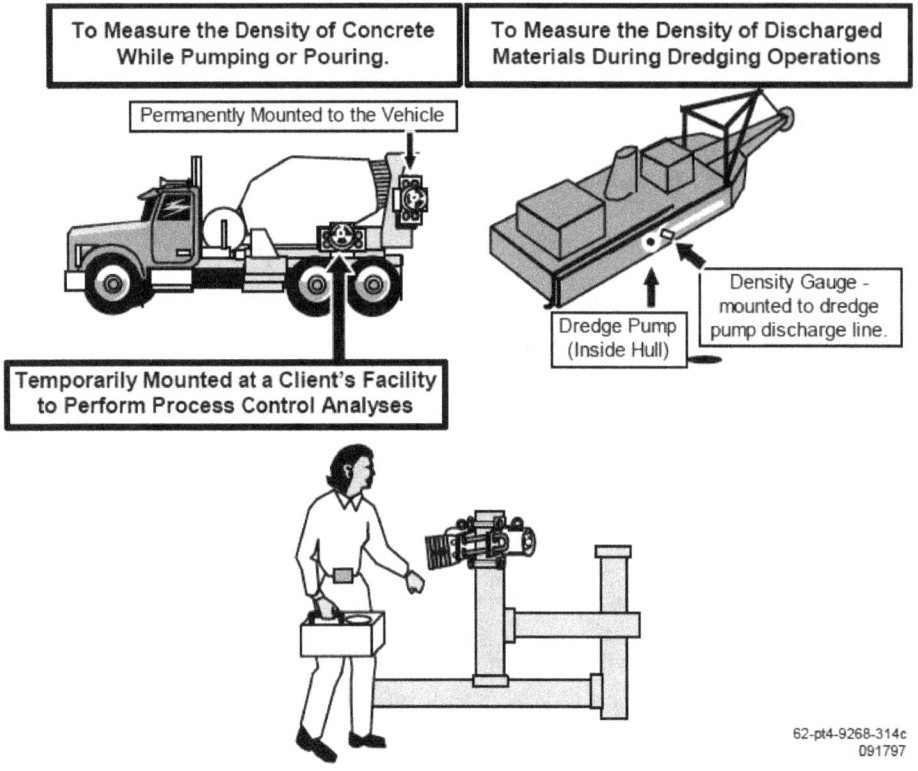

Figure 8.12 Examples of Uses for Fixed Gauges at Temporary Job Sites.

There are two categories of fixed gauges used at temporary job sites: Gauges that are permanently mounted to vehicles or trailers, and gauges that are transported to plants or refineries and temporarily installed on process equipment to conduct short-term QA/QC studies. See Figure 8.12.

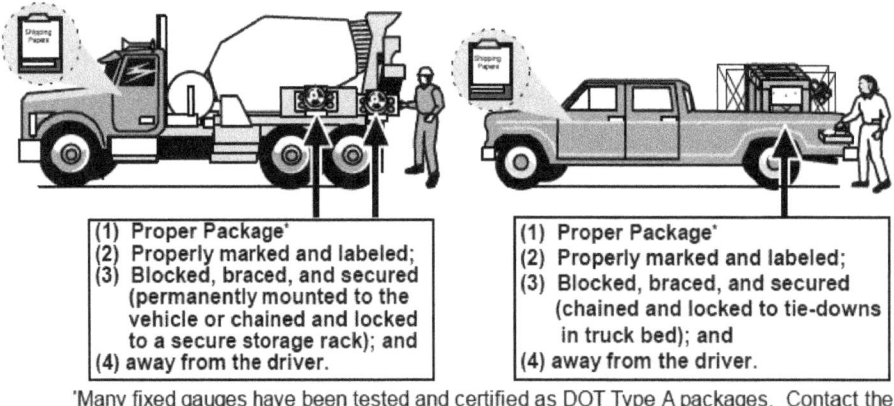

Figure 8.13 DOT Transportation Requirements.

Applicants must develop, implement, and maintain safety procedures for off-site transport of radioactive material to ensure compliance with DOT regulations. Figure 8.13 illustrates some important DOT requirements for gauge licensees. During an inspection, NRC uses the provisions of 10 CFR 71.5 and an MOU with DOT to examine and enforce transportation requirements applicable to fixed gauge licensees. Appendix O lists major DOT regulations and provides examples of shipping documents, placards, and labels.

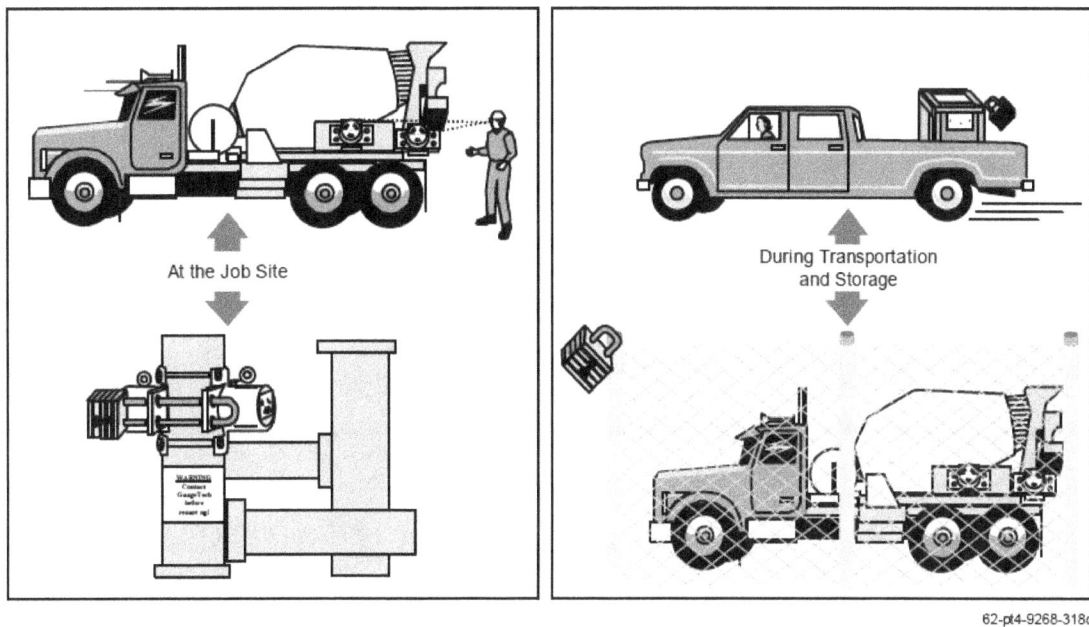

62-pt4-9268-318d
092497

Figure 8.14 Security. *Examples of Methods used to Secure Fixed Gauges at Temporary Job Sites.*

When working at a temporary job site, licensees generally have to follow the rules and procedures of the organization that owns or controls the site. Thus, licensees may not be able restrict access to areas in the same manner that they could at their own facilities. Furthermore, non-licensee personnel may not be familiar with fixed gauges or radioactive material. Therefore, to avoid lost or stolen gauges and to prevent unnecessary radiation exposures to members of the public, licensees must keep gauges under constant surveillance, or secured against unauthorized use or removal. See Figure 8.14.

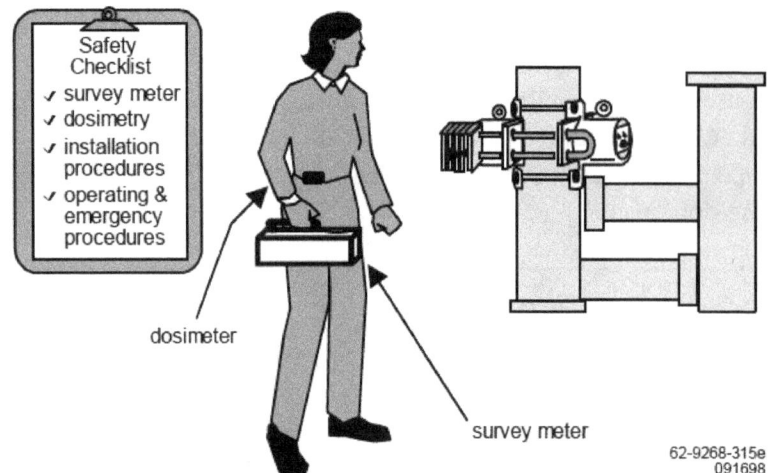

Figure 8.15 Installation of Fixed Gauges at Temporary Job Sites. *Examples of the Additional Precautions Needed when Installing Fixed Gauges at Temporary Job Sites.*

While installing gauges, personnel could receive radiation doses exceeding NRC limits if proper radiation safety principles are not followed. Licensee personnel performing installations should be assigned and wear personal dosimetry and use a survey meter to monitor radiological conditions. See Figure 8.15.

After installing a gauge at a temporary job site, a radiation survey should be conducted to ensure that dose rates in unrestricted areas will not exceed 0.02 mSv (2 mrem) in any one hour or 1 mSv (100 mrem) in a year. If surveys indicate that a member of the public (e.g., client personnel) could receive a dose exceeding these limits, licensees would need to adopt additional security measures to prevent public access such as maintaining constant surveillance or erecting physical barriers. See Figure 8.16.

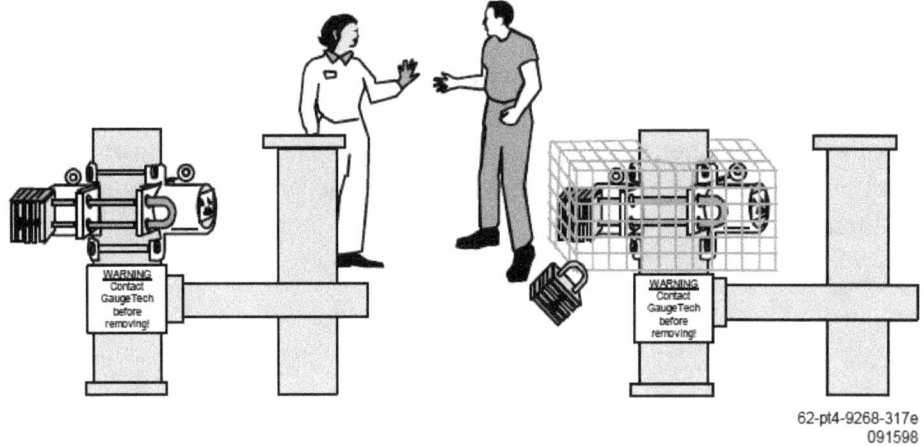

Figure 8.16 Security. *Additional Security Measures following Installation of Fixed Gauges at a Temporary Job Site.*

Response from Applicant: Submit one of the following three alternatives:

- A statement that: "We will not use fixed gauges at temporary job sites."

OR

- A statement that: "Procedures for use of fixed gauges at temporary job sites will be developed, implemented, maintained, and distributed and will meet the Criteria in the section entitled 'Radiation Safety Program - Fixed Gauges Used at Temporary Job Sites,' in NUREG-1556, Vol. 4, 'Consolidated Guidance about Materials Licenses: Program-Specific Guidance about Fixed Gauge Licenses,' dated October 1998 and copies of these procedures will be provided to all gauge users."

OR

- Alternative procedures for use of fixed gauges at temporary job sites.

Note: Alternative procedures will be evaluated using the criteria listed above.

8.10.11 MINIMIZATION OF CONTAMINATION

Regulations: 10 CFR 20.1406.

Criteria: Applicants for new licenses must describe how facility design and procedures for operation will minimize, to the extent practicable, contamination of the facility and the environment, facilitate eventual decommissioning, and minimize, to the extent practicable, the generation of radioactive waste.

Discussion: All applicants for new licenses need to consider the importance of designing and operating their facilities so as to minimize the amount of radioactive contamination generated at the site during its operating lifetime and to minimize the generation of radioactive waste during decontamination. In the case of fixed gauge applicants, these issues usually do not need to be addressed as a separate item, as they are included in responses to other items of the application.

Sealed sources and devices that are approved by NRC or an Agreement State and located and used according to the respective SS&D Registration Certificate usually pose little risk of contamination. Leak tests performed at the frequency specified in the SS&D Registration Certificate should identify defective sources. Leaking sources must be immediately withdrawn from use and decontaminated, repaired, or disposed of according to NRC requirements. These steps minimize the spread of contamination and reduce radioactive waste associated with decontamination efforts. Other efforts to minimize radioactive waste do not apply to programs using only sealed sources and devices that have not leaked.

Response from Applicant: The applicant does not need to provide a response to this item under the following condition. NRC will consider that the above criteria have been met if the applicant's responses meet the criteria for the following sections: Radioactive Material - Sealed Sources and Devices, Facilities and Equipment, Radiation Safety Program - Operating and Emergency Procedures, Radiation Safety Program - Leak Testing, and Waste Management - Gauge Transfer and Disposal.

8.11   ITEM 11: WASTE MANAGEMENT

8.11.1   GAUGE DISPOSAL AND TRANSFER

Regulations: 10 CFR 20.2001, 10 CFR 30.41, 10 CFR 30.51, 10 CFR 30.36.

Criteria: Licensed materials must be disposed of in accordance with NRC requirements by transfer to an authorized recipient. Appropriate records must be maintained.

Discussion: When disposing of fixed gauges, licensees must transfer them to an authorized recipient. Authorized recipients are the original manufacturer or distributor of the device, a commercial firm licensed by the NRC or an Agreement State to accept radioactive waste from other persons, or another specific licensee authorized to possess the licensed material (i.e., its license specifically authorizes the same radionuclide, form, and use).

Before transferring radioactive material, a licensee must verify that the recipient is properly authorized to receive it using one of the methods described in 10 CFR 30.41. In addition, all packages containing radioactive sources must be prepared and shipped in accordance with NRC and DOT regulations. Records of the transfer must be maintained as required by 10 CFR 30.51.

Response from Applicant: The applicant does not need to provide a response to this item during the licensing process. However, the licensee should establish and include waste disposal procedures in its radiation safety program.

Because of the difficulties and costs associated with disposal of sealed sources, applicants should preplan the disposal. Applicants may want to consider contractual arrangements with the source supplier as part of a purchase agreement. Significant problems can arise from improper gauge transfer or failure to dispose of gauges in a proper and timely manner. See IN 86-31, "Unauthorized Transfer and Loss of Control of Industrial Nuclear Gauges," dated May 5, 1986, and IN 88-02, "Lost or Stolen Gauges," dated February 2, 1988.

References: See the Notice of Availability (on the inside front cover of this report) to obtain copies of IN 86-31, "Unauthorized Transfer and Loss of Control of Industrial Nuclear Gauges," dated May 5, 1986 and IN 88-02, "Lost or Stolen Gauges," dated February, 2, 1988.

Items 12 and 13, discussed below, are to be completed on the form itself (NRC Form 313).

8.12   ITEM 12: FEES

On NRC Form 313, enter the appropriate fee category from 10 CFR 170.31 and the amount of the fee enclosed with the application.

Direct all questions about NRC's fees or completion of Item 12 of NRC Form 313 (Appendix A) to the Office of the Chief Financial Officer at NRC Headquarters in Rockville, MD, (301) 415-7554. You may also call NRC's toll free number, 800-368-5642, and ask for extension 415-7554.

8.13   ITEM 13: CERTIFICATION

Individuals acting in a private capacity are required to date and sign NRC Form 313. Otherwise, representatives of the corporation or legal entity filing the application should date and sign NRC Form 313. *Representatives signing an application must be authorized to make binding commitments and to sign official documents on behalf of the applicant.* As discussed previously in "Management Responsibility," signing the application acknowledges management's commitment and responsibilities for the radiation protection program. *NRC will return all unsigned applications for proper signature.*

Note:

- It is a criminal offense to make a willful false statement or representation on applications or correspondence (18 U.S.C. 1001).

- When the application references commitments, those items become part of the licensing conditions and regulatory requirements.

9   AMENDMENTS AND RENEWALS TO A LICENSE

It is the licensee's obligation to keep the license current. If any of the information provided in the original application is to be modified or changed, the licensee must submit an application for a license amendment before the change takes place. Also, to continue the license after its expiration date, the licensee must submit an application for a license renewal at least 30 days before the expiration date (10 CFR 2.109, 10 CFR 30.36(a)).

Applications for license amendment, in addition to the following, must provide the appropriate fee. For renewal and amendment requests applicants must do the following:

- Be sure to use the most recent guidance in preparing an amendment or renewal request.

- Submit in duplicate, either an NRC Form 313 or a letter requesting amendment or renewal.

- Provide the license number.

- For renewals, provide a complete and up-to-date application if many outdated documents are referenced or there have been significant changes in regulatory requirements, NRC's guidance, the licensee's organization, or radiation protection program. Alternately, describe clearly the exact nature of the changes, additions, and deletions.

> Using the suggested wording of responses and committing to using the model procedures in this report will expedite NRC's review.

10 APPLICATIONS FOR EXEMPTIONS

Various sections of NRC's regulations address requests for exemptions (e.g., 10 CFR 19.31, 10 CFR 20.2301, 10 CFR 30.11(a), 10 CFR 71.8). These regulations state that NRC may grant an exemption, acting on its own initiative or on an application from an interested person. Key considerations are whether the exemption is authorized by law, will endanger life or property or the common defense and security, and is otherwise in the public interest.

> Until NRC has granted an exemption in writing, NRC expects strict compliance with all applicable regulations.

Exemptions are not intended to revise regulations, are not intended for large classes of licenses, and are generally limited to unique situations. Exemption requests must be accompanied by descriptions of the following:

- Exemption and why it is needed

- Proposed compensatory safety measures intended to provide a level of health and safety equivalent to the regulation for which the exemption is being requested

- Alternative methods for complying with the regulation and why they are not feasible.

11 TERMINATION OF ACTIVITIES

Regulations: 10 CFR 20.1402, 10 CFR 20.1403, 10 CFR 30.34(b), 10 CFR 30.35(g), 10 CFR 30.36(d), 10 CFR 30.36(g), 10 CFR 30.36(h), 10 CFR 30.36(j), 10 CFR 30.51(f)

Criteria: The licensee must do the following:

- Notify NRC, in writing, within 60 days of:

  — the expiration of its license

  — a decision to permanently cease licensed activities at the *entire site* (regardless of contamination levels)

  — a decision to permanently cease licensed activities in *any separate building or outdoor area*, if they contain residual radioactivity making them unsuitable for release according to NRC requirements.

  — no principal activities having been conducted *at the entire site* under the license for a period of 24 months

  — no principal activities having been conducted for a period of 24 months in *any separate building or outdoor area*, if they contain residual radioactivity making them unsuitable for release according to NRC requirements.

- Submit decommissioning plan, if required by 10 CFR 30.36(g).

- Conduct decommissioning, as required by 10 CFR 30.36(h) and 10 CFR 30.36(j).

- Submit, to the appropriate NRC regional office, completed NRC Form 314, "Certificate of Disposition of Materials" (or equivalent information) and a demonstration that the premises are suitable for release for unrestricted use (e.g., results of final survey).

- Before a license is terminated, send the records important to decommissioning to the appropriate NRC regional office. If licensed activities are transferred or assigned in accordance with 10 CFR 30.34(b), transfer records important to decommissioning to the new licensee.

Discussion: As noted in several instances discussed in "Criteria," before a licensee can decide whether it must notify NRC, the licensee must determine whether residual radioactivity is present and if so, whether the levels make the building or outdoor area unsuitable for release according to NRC requirements. A licensee's determination that a facility is not contaminated is subject to verification by NRC inspection.

For guidance on the disposition of licensed material, see the section on "Waste Management - Gauge Disposal or Transfer." For guidance on decommissioning records, see the section on "Radioactive Materials - Financial Assurance and Record Keeping for Decommissioning."

Response from Applicant: The applicant is not required to submit a response to the NRC during the initial application. However, when the license expires or at the time the licensee ceases operations, then any necessary decommissioning activities must be undertaken and NRC Form 314 or equivalent information must be submitted, and other actions must be taken as summarized in criteria above.

References: Copies of NRC Form 314, "Certificate of Disposition of Materials," are available upon request from NRC's Regional Offices. See Figure 2.1 for addresses and telephone numbers.

Appendix A

NRC Form 313

NRC Form 313

Replace this page with NRC Form 313

Appendix B

Suggested Format for Providing Information Requested in Items 5 Through 11 of NRC Form 313

Suggested Format for Providing Information Requested in Items 5 Through 11 of NRC Form 313

Table B.1 Items 5 & 6: Materials To Be Possessed and Proposed Uses

Yes	No	Radioisotope	Manufacturer or Distributor Model No.	Quantity	Use As Listed on SSD Certificate	Specify Other Uses Not Listed on SSD Certificate
		Cobalt-60	Sealed source manufacturer or distributor and model number: _____ Device manufacturer or distributor and model number: _____	Not to exceed either the maximum activity per source or maximum activity per device as specified in Sealed Source and Device Registration Certificate	Yes [ ] Specific description of the gauge use: _____	[ ] Not applicable _____ [ ] Uses are: _____ (Submit safety analysis supporting safe use)
		Krypton-85	Sealed source manufacturer or distributor and model number: _____ Device manufacturer or distributor and model number: _____	Not to exceed either the maximum activity per source or maximum activity per device as specified in Sealed Source and Device Registration Certificate	Yes [ ] Specific description of the gauge use: _____	[ ] Not applicable _____ [ ] Uses are: _____ (Submit safety analysis supporting safe use)
		Strontium-90	Sealed source manufacturer or distributor and model number: _____ Device manufacturer or distributor and model number: _____	Not to exceed either the maximum activity per source or maximum activity per device as specified in Sealed Source and Device Registration Certificate	Yes [ ] Specific description of the gauge use: _____	[ ] Not applicable _____ [ ] Uses are: _____ (Submit safety analysis supporting safe use)
		Cesium-137	Sealed source manufacturer or distributor and model number: _____ Device manufacturer or distributor and model number: _____	Not to exceed either the maximum activity per source or maximum activity per device as specified in Sealed Source and Device Registration Certificate	Yes [ ] Specific description of the gauge use: _____	[ ] Not applicable _____ [ ] Uses are: _____ (Submit safety analysis supporting safe use)

APPENDIX B

Yes	No	Radioisotope	Manufacturer or Distributor Model No.	Quantity	Use As Listed on SSD Certificate	Specify Other Uses Not Listed on SSD Certificate
		Americium-241	Sealed source manufacturer or distributor and model number: Device manufacturer or distributor and model number:	Not to exceed either the maximum activity per source or maximum activity per device as specified in Sealed Source and Device Registration Certificate	Yes [ ] Specific description of the gauge use: _____ _____ _____ _____ _____	[ ] Not applicable _____ [ ] Uses are: _____ (Submit safety analysis supporting safe use)
		Other Isotope (Specify):	Sealed source manufacturer or distributor and model number: Device manufacturer or distributor and model number:	Not to exceed either the maximum activity per source or maximum activity per device as specified in Sealed Source and Device Registration Certificate	Yes [ ] Specific description of the gauge use: _____ _____ _____ _____ _____	[ ] Not applicable _____ [ ] Uses are: _____ (Submit safety analysis supporting safe use)
		*Financial Assurance Required **and Evidence of Financial Assurance Provided***				

Table B.2   Items 7 Through 11: Training and Experience, Facilities and Equipment, Radiation Safety Program, and Waste Disposal

Item No. and Title	Suggested Response	Yes	Alternative Procedures Attached
7. Individual(s) Responsible For Radiation Safety Program And Their Training And Experience 7.1 Radiation Safety Officer Name: _____	Before obtaining licensed materials, the proposed RSO will have successfully completed the training described in Criteria in the section entitled "Individual(s) Responsible for Radiation Safety Program and Their Training and Experience - Radiation Safety Officer" in NUREG-1556, Vol. 4, dated October 1998. **AND** Before being named as the RSO, future RSOs will have successfully completed the training described in Criteria in the section entitled "Individual(s) Responsible for Radiation Safety Program and Their Training and Experience - Radiation Safety Officer" in NUREG-1556, Vol. 4, dated October 1998. Within 30 days of naming a new RSO, we will submit the new RSO's name to NRC to include in our license.	[ ]	[ ]
7. Individual(s) Responsible For Radiation Safety Program And Their Training And Experience 7.2 Authorized Users	PROPOSED AUTHORIZED USERS: Before using licensed materials, authorized users will have successfully completed the training described in Criteria in the section entitled, "Authorized Users" in NUREG-1556, Vol. 4, dated October 1998.	[ ]	[ ]
8. Training for Individuals Who in the Course of Employment are Likely to Receive Occupational Doses of Radiation in Excess of 1 mSv (100 mrem) in a Year (Occupationally Exposed Workers) and Ancillary Personnel	The applicant is *not* required to, and should not, submit is training program, for individuals who in the course of employment are likely to receive occupational doses of radiation in excess of 1 mSv (100 mrem) in a year (occupationally exposed workers) and ancillary personnel, to the NRC for review during the licensing phase.	Need Not Be Submitted with Application	

Item No. and Title	Suggested Response	Yes	Alternative Procedures Attached
9. Facilities and Equipment	We will ensure that the location of each fixed gauge meets the Criteria in the section entitled "Facilities and Equipment" in NUREG-1556, Vol. 4, dated October 1998.	[ ]	[ ]
10. Radiation Safety Program - Audit Program	The applicant is *not* required to, and should not, submit its audit program to the NRC for review during the licensing phase.	Need Not Be Submitted with Application	
10. Radiation Safety Program - Survey Instruments	Surveys pursuant to 10 CFR 20.1501 will be performed by a person specifically authorized by the NRC or an Agreement State to perform these surveys. **OR** We will use instruments that meet the Criteria in the section entitled "Radiation Safety Program - Instruments," in NUREG-1556, Vol. 4, dated August 1998, and *one* of the following: Each survey meter will be calibrated by the manufacturer or other person authorized by the NRC or an Agreement State to perform survey meter calibrations. **OR** We will implement the model survey instrument calibration program in Appendix I to NUREG-1556, Vol. 4, dated October 1998.	[ ]	[ ]
10. Radiation Safety Program - Material Receipt and Accountability	Physical inventories will be conducted at intervals not to exceed 6 months or at other intervals approved by the NRC, to account for all sealed sources and devices received and possessed under the license.	[ ]	[ ]
10. Radiation Safety Program - Occupational Dosimetry	We will perform a prospective evaluation demonstrating that unmonitored individuals are not likely to receive, in one year, a radiation dose in excess of 10% of the allowable limits in 10 CFR Part 20 or we will provide dosimetry that meets the Criteria in the section entitled "Radiation Safety Program - Occupational Dosimetry," in NUREG-1556, Vol. 4, dated October 1998.	[ ]	[ ]

Item No. and Title	Suggested Response	Yes	Alternative Procedures Attached
10. Radiation Safety Program - Public Dose	The applicant is not required to submit a response to the public dose section during the licensing phase. However, during NRC inspections, licensees must be able to provide documentation demonstrating, by measurement or calculation, that the total effective dose equivalent to the individual likely to receive the highest dose from the licensed operation does not exceed the annual limit for individual members of the public.	Need Not Be Submitted with Application	
10. Radiation Safety Program - Operating & Emergency Procedures	If the gauge meets one or more of the safety conditions specified in "Discussion," in the section entitled "Radiation Safety Program-Operating Emergency Procedures," in NUREG 1556, Vol. 4, dated August 1998 state the following: Operating and emergency procedures will be developed, implemented, maintained, and distributed, and will meet the Criteria in the section entitled "Radiation Safety Program - Operating and Emergency Procedures," in NUREG-1556, Vol. 4, dated August 1998. For each gauge requested that does not meet one or more of the safety conditions specified in "Discussion," in the section entitled "Radiation Safety Program-Operating Emergency Procedures," in NUREG 1556, Vol. 4, dated August 1998 provide your operating, emergency and lock-out (if applicable) procedures to NRC for review.	[ ] [ ] Procedures Attached	[ ]
10. Radiation Safety Program - Leak Test	Leak tests will be performed at intervals approved by the NRC or an Agreement State and specified in the Sealed Source and Device Registration Certificate. Leak tests will be performed by an organization authorized by NRC or an Agreement State to provide leak testing services for other licensees or using a leak test kit supplied by an organization authorized by NRC or an Agreement State to provide leak test kits to other licensees and according to the kit supplier's instructions. **OR** We will implement the model leak test program published in Appendix M to NUREG-1556, Vol. 4, dated October 1998.	[ ] [ ]	[ ]

Item No. and Title	Suggested Response	Yes	Alternative Procedures Attached
10. Radiation Safety Program - Maintenance	ROUTINE MAINTENANCE We will implement and maintain procedures for routine maintenance of our fixed gauges according to each manufacturer's or distributor's written recommendations and instructions. NON-ROUTINE MAINTENANCE OPERATIONS The gauge manufacturer, distributor or other person authorized by NRC or an Agreement State will perform non-routine operations such as installation, initial radiation survey, repair, and maintenance of components related to the radiological safety of the gauge, gauge relocation, replacement, and disposal of sealed sources, alignment, or removal of a gauge from service.	 [ ]	[ ] [ ] The information listed in Appendix N supporting a request to perform non-routing operations in-house is attached
10. Radiation Safety Program - Transportation	The applicant is *not* required to submit its response to transportation during the licensing process; this issue will be reviewed during inspection. However, the licensee should develop, implement, and maintain transportation procedures according to NRC and DOT regulations.	Need Not Be Submitted with Application	
10. Radiation Safety Program - Fixed Gauges Used at Temporary Job Sites .	This is not applicable to our program. We will not use fixed gauges at temporary job sites. **OR** We will develop, implement, maintain and distribute procedures that meet the Criteria in the section entitled "Radiation Safety Program - Fixed Gauges Used at Temporary Job Sites" in NUREG-1556, Vol. 4, dated October 1998.	[ ] Not Applicable [ ]	 [ ]
10. Radiation Safety Program - Minimization of Contamination	The applicant is not required to submit a response to minimization of contamination if the applicant's responses meet the criteria for the following sections: Radioactive Material - Sealed Sources and Devices, Facilities and Equipment, Radiation Safety Program - Operating and Emergency Procedures, Radiation Safety Program - Leak Testing, and Waste Management - Gauge Transfer and Disposal.	Need Not Be Submitted with Application	

Item No. and Title	Suggested Response	Yes	Alternative Procedures Attached
11. Waste Management - Gauge Disposal & Transfer	The applicant is not required to submit a response to waste management during the licensing process. However, the licensee should develop, implement, and maintain gauge transfer and disposal procedures in its radiation protection program.	Need Not Be Submitted with Application	

Appendix C

Information Needed for Transfer of Control Application

Information Needed for Transfer of Control Application

Licensees must provide full information and obtain NRC's *prior written consent* before transferring control of the license; some licensees refer to this as "transferring the license." Provide the following information concerning changes of control by the applicant (transferor and/or transferee, as appropriate). If any items are not applicable, so state.

1. The new name of the licensed organization. If there is no change, the licensee should so state.

2. The new licensee contact and telephone number(s) to facilitate communications.

3. Any changes in personnel having control over licensed activities (e.g., officers of a corporation) and any changes in personnel named in the license such as radiation safety officer, authorized users, or any other persons identified in previous license applications as responsible for radiation safety or use of licensed material. The licensee should include information concerning the qualifications, training, and responsibilities of new individuals.

4. An indication of whether the transferor will remain in non-licensed business without the license.

5. A complete, clear description of the transaction, including any transfer of stocks or assets, mergers, etc., so that legal counsel is able, when necessary, to differentiate between name changes and transferring control.

6. A complete description of any planned changes in organization, location, facility, equipment, or procedures (i.e., changes in operating or emergency procedures).

7. A detailed description of any changes in the use, possession, location, or storage of the licensed materials.

8. Any changes in organization, location, facilities, equipment, procedures, or personnel that would require a license amendment even without transferring control.

9. An indication of whether all surveillance items and records (e.g., calibrations, leak tests, surveys, inventories, and accountability requirements) will be current at the time of transfer. Provide a description of the status of all surveillance requirements and records.

10. Confirmation that all records concerning the safe and effective decommissioning of the facility, pursuant to 10 CFR 30.35(g), 40.36(f), 70.25(g), and 72.30(d); public dose; and waste disposal by release to sewers, incineration, radioactive material spills, and on-site burials, have been transferred to the new licensee, if licensed activities will continue at the same location, or to the NRC for license terminations.

11. A description of the status of the facility. Specifically, the presence or absence of contamination should be documented. If contamination is present, will decontamination occur before transfer? If not, does the successor company agree to assume full liability for the decontamination of the facility or site?

12. A description of any decontamination plans, including financial assurance arrangements of the transferee, as specified in 10 CFR 30.35, 40.36, and 70.25. Include information about how the transferee and transferor propose to divide the transferor's assets, and responsibility for any cleanup needed at the time of transfer.

13. Confirmation that the transferee agrees to abide by all commitments and representations previously made to NRC by the transferor. These include, but are not limited to: maintaining decommissioning records required by 10 CFR 30.35(g); implementing decontamination activities and decommissioning of the site; and completing corrective actions for open inspection items and enforcement actions.

    With regard to contamination of facilities and equipment, the transferee should confirm, in writing, that it accepts full liability for the site, and should provide evidence of adequate resources to fund decommissioning; or the transferor should provide a commitment to decontaminate the facility before transferring control.

    With regard to open inspection items, etc., the transferee should confirm, in writing, that it accepts full responsibility for open inspection items and/or any resulting enforcement actions; or the transferee proposes alternative measures for meeting the requirements; or the transferor provides a commitment to close out all such actions with NRC before license transfer.

14. Documentation that the transferor and transferee agree to transfer control of the licensed material and activity, and the conditions of transfer; and the transferee is made aware of all open inspection items and its responsibility for possible resulting enforcement actions.

15. A commitment by the transferee to abide by all constraints, conditions, requirements, representations, and commitments identified in the existing license. If not, the transferee must provide a description of its program, to ensure compliance with the license and regulations.

References: The information above is contained in IN 89-25, Revision 1, "Unauthorized Transfer of Ownership or Control of Licensed Activities." See the Notice of Availability (on the inside front cover of this report) to obtain copies.

Appendix D

Reviewer Checklist for Fixed Gauge Application

Reviewer Checklist for Fixed Gauge Application

ITEM 1: ACTION TYPE

ACTION TYPE:	ADMINISTRATIVE REVIEW:
[ ] New [ ] Amendment [ ] Renewal	[ ] Current Guidance Used [ ] References in Application Based On Current Regulations [ ] All Attachments Referenced Included [ ] Signature on Application

ITEM 2: **LEGAL IDENTITY**

NAME:	

ITEMS 2 & 3: **ADDRESS**

LOCATION OF USE/STORAGE ADDRESS:	MAILING ADDRESS:

ITEM 4: **PERSON TO BE CONTACTED ABOUT THIS APPLICATION**

CONTACT PERSON:	
TELEPHONE NUMBER:	

Table D.1   Items 5 and 6: Materials to Be Possessed and Uses

Yes	No	Radioisotope	Model No.	Quantity	Use As Listed on SSD Certificate	Specify Other Uses Not Listed on SSD Certificate
		Cobalt-60	Sealed source manufacturer or distributor and model number: _____ Device manufacturer or distributor and model number:	Not to exceed either the maximum activity per source or maximum activity per device as specified in Sealed Source and Device Registration Certificate	Yes [ ] Specific description of the gauge use: _____ _____ _____ _____ _____	[ ] Not applicable _____ [ ] Uses are:
		Krypton-85	Sealed source manufacturer or distributor and model number: _____ Device manufacturer or distributor and model number:	Not to exceed either the maximum activity per source or maximum activity per device as specified in Sealed Source and Device Registration Certificate	Yes [ ] Specific description of the gauge use: _____ _____ _____ _____ _____	[ ] Not applicable _____ [ ] Uses are:
		Strontium-90	Sealed source manufacturer or distributor and model number: _____ Device manufacturer or distributor and model number:	Not to exceed either the maximum activity per source or maximum activity per device as specified in Sealed Source and Device Registration Certificate	Yes [ ] Specific description of the gauge use: _____ _____ _____ _____ _____	[ ] Not applicable _____ [ ] Uses are:
		Cesium-137	Sealed source manufacturer or distributor and model number: _____ Device manufacturer or distributor and model number:	Not to exceed either the maximum activity per source or maximum activity per device as specified in Sealed Source and Device Registration Certificate	Yes [ ] Specific description of the gauge use: _____ _____ _____ _____ _____	[ ] Not applicable _____ [ ] Uses are:

Yes	No	Radioisotope	Model No.	Quantity	Use As Listed on SSD Certificate	Specify Other Uses Not Listed on SSD Certificate
		Americium-241	Sealed source manufacturer or distributor and model number: _____ Device manufacturer or distributor and model number: _____	Not to exceed either the maximum activity per source or maximum activity per device as specified in Sealed Source and Device Registration Certificate	Yes [ ] Specific description of the gauge use: _____	[ ] Not applicable _____ [ ] Uses are:
		Other Isotope (Specify):	Sealed source manufacturer or distributor and model number: _____ Device manufacturer or distributor and model number: _____	Not to exceed either the maximum activity per source or maximum activity per device as specified in Sealed Source and Device Registration Certificate	Yes [ ] Specific description of the gauge use: _____	[ ] Not applicable _____ [ ] Uses are:
		*Financial Assurance Required **and Evidence of Financial Assurance Provided***				

Table D.2   Items 7 Through 11: Training and Experience, Facilities and Equipment, Radiation Safety Program, and Waste Management

Item Number and Title	Suggested Response	Applicant's Response			
		Yes	No	Other	
				Yes	No
7. **Individual(s) Responsible for Radiation Safety Program and their Training and Experience** **7.1 Radiation Safety Officer (RSO)** **Name:** _____	Before obtaining licensed materials, the proposed RSO will have successfully completed the training described in Criteria in the section entitled "Radiation Safety Officer," in NUREG-1556, Vol. 4 dated August 1998. **AND** Before being named as the RSO, future RSOs will have successfully completed the training described in Criteria in the section entitled "Radiation Safety Officer," in NUREG-1556, Vol. 4, dated August 1998. Within 30 days of naming a new RSO, we will submit the new RSO's name to NRC to include in our license.				

Item Number and Title	Suggested Response	Applicant's Response			
		Yes	No	Other	
				Yes	No

7. **Individual(s) Responsible for Radiation Safety Program and their Training and Experience**
 	
7.1 **Radiation Safety Officer (RSO)**
(Cont'd)
	Optional Response
	
	Criteria for Acceptable Training Course for Radiation Safety Officer
	
	Classroom Training:
	
	• Radiation Safety
	  – Radiation vs. contamination
	  – Internal vs. external exposure
	  – Biological effects of radiation
	  – Types and relative hazards of radioactive material possessed
	  – ALARA concept
	  – Use of time, distance, and shielding to minimize exposure
	  – Locations of sealed source within the gauge
	  – Use of survey meters and personal dosimetry, when required
	
	• Regulatory Requirements
	  – Applicable regulations
	  – License conditions, amendments, renewals
	  – Locations of use and storage of radioactive materials
	  – Material control and accountability
	  – Annual audit of radiation safety program
	  – Transfer and disposal
	  – Recordkeeping
	  – Prior events involving fixed gauges
	  – Handling incidents
	  – Recognizing and ensuring that radiation warning signs are visible and legible
	  – Licensing and inspection by regulatory agency
	  – Need for complete and accurate information
	  – Employee protection
	  – Deliberate misconduct
	
	• Practical Explanation of the Theory and Operation for Each Gauge Possessed by the Licensee
	  – Operating and emergency procedures
	  – Routine vs. non-routine maintenance
	  – Lock-out procedures				

Item Number and Title	Suggested Response	Applicant's Response			
		Yes	No	Other	
				Yes	No
7. Individual(s) Responsible for Radiation Safety Program and their Training and Experience **7.1 Radiation Safety Officer (RSO)** *(Cont'd)*	• Supervised "Hands-On" Experience performing – Operating procedures – Test runs of emergency procedures – Routine maintenance – Lock-out procedures **Training Assessment** Course Instructor Qualifications: • Bachelor's degree in a physical or life science or engineering with successful completion of both a fixed gauge manufacturer's or distributor's course for users and an 8 hour radiation safety course and 8 hours hands-on experience with fixed gauges **OR** • Successful completion of a fixed gauge manufacturer's or distributor's course for users • Successful completion of 40 hour radiation safety course • 30 hours of hands-on experience with fixed gauges. *Note:* Additional training is required for those applicants intending to perform non-routine operations.				
7 Individual(s) Responsible for Radiation Safety Program and their Training and Experience **7.2 Authorized Users**	**Proposed Authorized Users** Before using licensed materials, authorized users will have successfully completed the training described in Criteria in the section entitled "Authorized Users," in NUREG-1556, Vol. 4, dated August 1998.				

Item Number and Title	Suggested Response	Applicant's Response			
		Yes	No	Other	
				Yes	No

7 **Individual(s) Responsible for Radiation Safety Program and their Training and Experience** 7.2 **Authorized Users** *(Cont'd)*	**Optional Response** Classroom Training:
	
	• Radiation Safety
	  – Radiation vs. contamination
	  – Internal vs. external exposure
	  – Biological effects of radiation
	  – Types and relative hazards of radioactive material possessed
	  – ALARA concept
	  – Use of time, distance, and shielding to minimize exposure
	  – Location of sealed source within the gauge
	  – Use of survey meters and personal dosimetry, when required
	
	• Regulatory Requirements
	  – Applicable regulations
	  – License conditions, amendments, renewals
	  – Locations of use and storage of radioactive materials
	  – Material control and accountability
	  – Annual audit of radiation safety program
	  – Transfer and disposal
	  – Recordkeeping
	  – Prior events involving fixed gauges
	  – Handling incidents
	  – Recognizing and ensuring that radiation warning signs are visible and legible
	  – Licensing and inspection by regulatory agency
	  – Need for complete and accurate information
	  – Employee protection
	  – Deliberate misconduct
	
	• Practical Explanation of the Theory and Operation for Each Type of Gauge that may be used by the Authorized User
	  – Operating and emergency procedures
	  – Routine vs. non-routine maintenance
	  – Lock-out procedures				

Item Number and Title	Suggested Response	Applicant's Response			
		Yes	No	Other	
				Yes	No

	• Supervised Hands-on Experience Performing
	  – Operating procedures
	  – Test runs of emergency procedures
	  – Routine maintenance
	  – Lock-out procedures				

7 **Individual(s) Responsible for Radiation Safety Program and their Training and Experience** **7.2 Authorized Users** *(Cont'd)*	**Training Assessment** Course Instructor Qualifications: • Bachelor's degree in a physical or life science or engineering with successful completion of both a fixed gauge manufacturer's or distributor's course for users and an 8 hour radiation safety course and 8 hours hands-on experience with fixed gauges **OR** • Successful completion of a fixed gauge manufacturer's or distributor's course for users • Successful completion of 40 hour radiation safety course • 30 hours of hands-on experience with fixed gauges *Note:* • Individuals who in the course of employment are likely to receive occupational doses of radiation in excess of 1 mSv (100 mrem) in a year must receive training pursuant to 10 CFR 19.12. • Additional training is required for those applicants requesting to perform non-routine operations.				
8 **Training for Individuals Who in the Course of Employment are Likely to Receive Occupational Doses of Radiation in Excess of 1 mSv (100 mrem) in a Year (Occupationally Exposed Workers) and Ancillary Personnel**	The applicant is not required to, and should not, submit its training program, for individuals who in the course of employment are likely to receive occupational doses of radiation in excess of 1 mSv (100 mrem) in a year (occupationally exposed workers) and ancillary personnel, to the NRC for review during the licensing phase.	Need Not Be Submitted with Application			

Item Number and Title	Suggested Response	Applicant's Response			
		Yes	No	Other	
				Yes	No
9 Facilities and Equipment	We will ensure that the location of each fixed gauge meets the Criteria in section entitled "Facilities and Equipment," in NUREG-1556, Vol. 4, dated August 1998. **OR** Confirm that the fixed gauge is secured to prevent unauthorized removal or access; and submit specific information supporting the new conditions demonstrating that they will not impact the safety or integrity of the source or device. Address any instances where the proposed conditions exceed any conditions listed in the SSD Registration Certificate **Optional Response** • The area corresponds to the "Conditions of Normal Use" and "Limitations and/or Other Considerations of Use" on the SSD Registration Certificate • The fixed gauge is secured to prevent unauthorized removal (e.g., located in a locked room, permanently mounted, or chained and locked to a storage rack)				
10 Radiation Safety Program - Audit Program	The applicant is not required to, and should not, submit its audit program to the NRC for review during the licensing phase	Need Not Be Submitted with Application			

Item Number and Title	Suggested Response	Applicant's Response			
		Yes	No	Other	
				Yes	No
10 Radiation Safety Program - Instruments	Surveys pursuant to 10 CFR 20.1501 will be performed by a person specifically authorized by the NRC or an Agreement State to perform these surveys." **OR** We will use instruments that meet the Criteria in the section entitled "Radiation Safety Program - Instruments," in NUREG-1556, Vol. 4, dated August 1998, and one of the following: • Each survey meter will be calibrated by the manufacturer or other person authorized by the NRC or an Agreement State to perform survey meter calibrations. **OR** • We will follow the model survey instrument calibration program in Appendix I to NUREG-1556, Vol. 4, dated August 1998. **Optional Response** The applicant may provide a description of an alternate method to perform surveys pursuant to 10 CFR 20.1501.				
10 Radiation Safety Program - Instrument Calibration	If required to do surveys pursuant to 10 CFR 20.1501, and requesting to calibrate their own survey meters: We will implement the model survey instrument calibration program published in Appendix I to NUREG - 1556, Vol. 4, dated October 1998.				

APPENDIX D

Item Number and Title	Suggested Response	Applicant's Response			
		Yes	No	Other	
				Yes	No

10 Radiation Safety Program - Instrument Calibration *(Cont'd)*	**Optional Response**
	
	• Training and experience of individual performing calibration.
	• Description of facilities, equipment
	• Specify calibration source radionuclide, activity, traceability (source activity sufficient to provide a dose rate of at least 30 mR/hr at 100 cm, similar in energy to gauge sources. NIST traceable)
	• Specific procedures for calibration
	• Calibration report
	• Calibration tag, sticker:
	  – source
	  – for each scale or decade not calibrated, indication checked for function only or scale not operative
	  – calibration date
	  – due date
	  – exposure rate from check source if used
	
	• Maintain calibration records for 3 years				

10 Radiation Safety Program - Material Receipt and Accountability	Physical inventories will be conducted at intervals not to exceed 6 months or at other intervals as approved by the NRC, to account for all sealed sources and devices received and possessed under the license. **Optional Response** A description of the procedures for ensuring that no fixed gauge has been lost, stolen, or misplaced and how often they will be conducted.				
10 Radiation Safety Program - Occupational Dosimetry	We will perform a prospective evaluation demonstrating that unmonitored individuals are not likely to receive, in one year, a radiation dose in excess of 10% of the allowable limits in 10 CFR Part 20 or we will provide dosimetry that meets the Criteria in the section entitled "Radiation Safety Program - Occupational Dosimetry," in NUREG-1556, Vol. 4, dated October 1998. **Optional Response** Alternative response demonstrates compliance with 10 CFR Part 20 requirements.				

Item Number and Title	Suggested Response	Applicant's Response			
		Yes	No	Other	
				Yes	No
10 **Radiation Safety Program - Public Dose**	The applicant is not required to submit a response to public dose section during the licensing phase. Documentation demonstrating compliance will be examined during inspection.	Need Not Be Submitted with Application			
10 **Radiation Safety Program - Operating & Emergency Procedures**	If the gauge meets one or more of the safety conditions specified in "Discussion," in the section entitled "Radiation Safety Program - Operating Emergency Procedures," in NUREG 1556, Vol. 4, dated October 1998 state the following: Operating and emergency procedures will be developed, implemented and maintained and will meet the Criteria in the section entitled "Radiation Safety Program - Operating and Emergency Procedures," in NUREG-1556, Vol. 4, dated October 1998. For each gauge requested that does not meet one or more of the safety conditions specified in "Discussion," in the section entitled "Radiation Safety Program - Operating Emergency Procedures," in NUREG 1556, Vol. 4, dated October 1998 provide your operating, emergency and lock-out (if applicable) procedures to NRC for review.				

Item Number and Title	Suggested Response	Applicant's Response			
		Yes	No	Other	
				Yes	No
10 **Radiation Safety Program - Operating & Emergency Procedures** *(Cont'd)*	**Optional Response** For each type of gauge: • Operating Procedures – Instructions for operating the gauge – Instructions for performing routine cleaning and maintenance according to the manufacturers' or distributors' recommendations and instructions – Instructions for testing each gauge for the proper operation of the on/off mechanism (shutter) and indicator, if any, at intervals not to exceed 6 months or as specified in the SSD certificate – Instructions for lock-out procedures, if applicable, that are adequate to ensure that no individual or portion of an individual's body can enter the radiation beam. – Instructions to prevent unauthorized access, removal, or use of fixed gauges – Steps to take to keep radiation exposures ALARA – Steps to maintain accountability (i.e., physical inventory) – Instructions to ensure that non-routine operations such as installation, initial radiation survey, repair, and maintenance of components related to the radiological safety of the gauge, gauge relocation, replacement and disposal of sealed sources, alignment, or removal of a gauge from service are performed by the manufacturer, distributor or person specifically authorized by the NRC or an Agreement State – Steps to ensure that radiation warning signs are present, visible, and legible				

Item Number and Title	Suggested Response	Applicant's Response			
		Yes	No	Other	
				Yes	No
10 Radiation Safety Program - Operating & Emergency Procedures *(Cont'd)*	Emergency Procedures: • Stop use of the gauge • Restrict access to the area • Contact responsible and individuals (Telephone numbers for the RSO, authorized users, the gauge manufacturer or distributor, fire department, or other emergency response organization, as appropriate, and the NRC should be posted or easily accessible) • Do not attempt repair or authorize others to attempt repair of the gauge except as specifically authorized in a license issued by the NRC or an Agreement State • Require reporting to NRC pursuant to 10 CFR 20.2201-20.2203, 10 CFR 30.50, and 10 CFR 21.21 • Take additional steps, dependent on the specific situations. *Note:* • Copies of operating and emergency procedures provided to all gauge users • Post copies of operating and emergency procedures at each location of use or post a notice describing where procedures may be examined.				
10 Radiation Safety Program - Leak Tests	• Leak tests will be performed at intervals approved by the NRC or an Agreement State and Specified in the SSD Registration Certificate. Leak tests will be performed by an organization authorized by NRC or an Agreement State to provide leak testing services for other licensees or using a leak test kit supplied by an organization authorized by NRC or an Agreement State to provide leak test kits to other licensees and according to the kit supplier's instructions. Records of leak test results will be maintained. **OR** • We will implement the model leak test program published in Appendix M to NUREG-1556, Vol. 4, dated October 1998.				

Item Number and Title	Suggested Response	Applicant's Response			
		Yes	No	Other	
				Yes	No
10 **Radiation Safety Program - Leak Tests**	**Optional Response** • Identify the individual who will make the analysis; their training and experience • Leak test frequency as specified in the appropriate Sealed Source and Device Registration Certificate. • How and where test samples taken; materials to be used; methods of handling samples to prevent or minimize exposure to personnel. • Type of instrument(s) used, counting efficiency, and minimum levels of detection for each radionuclide *Note:* An instrument capable of making quantitative measurements should be used; hand-held survey meters will not normally be considered adequate for measurements. • Specify the standard calibration sources including for each: the radionuclide, quantity, accuracy, and traceability to primary radiation standards *Note:* Accuracy of standards should be within ± 5% of the stated value and traceable to a primary radiation standard such as those maintained by the National Institutes of Standards and Technology (NIST). • Sample calculation to convert measurement data to becquerels (or microcuries) • Instructions on actions, notifications regarding leaking source • Maintain records of leak test results				
10 **Radiation Safety Program - Maintenance**	**Routine Maintenance** We will implement and maintain procedures for routine maintenance of our gauges according to each manufacturer's or distributor's written recommendations and instructions. **Optional Response** • Adequate training, experience • Manufacturer's or distributor's written instructions • Considers ALARA • Ensures gauge functions as designed • Ensures source integrity not compromised				

Item Number and Title	Suggested Response	Applicant's Response			
		Yes	No	Other	
				Yes	No

10 Radiation Safety Program - Maintenance *(Cont'd)*	**Non-Routine Operations**
	
	The gauge manufacturer, distributor or other person authorized by NRC or an Agreement State will perform non-routine operations such as installation, initial radiation survey, repair, and maintenance of components related to the radiological safety of the gauge, gauge relocation, replacement, and disposal of sealed sources, alignment, or removal of a gauge from service.
	
	Optional Response
	
	Provide the information listed in Appendix N supporting a request to perform non-routine operations in-house.
	
	• Types of work to be performed
	• Identify the individual who will perform non-routine operations, their training and experience
	• Procedures to ensure:
	  – doses to public, personnel are ALARA and within regulatory limits
	  – security
	  – posting
	  – manufacturers or distributors instructions and recommendations are followed
	  – non-manufacturer/non-distributor supplied replacement components or parts, or the use of materials (e.g., lubricants) other than those specified or recommended by the manufacturer or distributor are evaluated to ensure that they do not degrade the engineering safety analysis
	  – before being returned to routine use, the gauge is tested to verify that it functions as designed and source integrity is not compromised
	• Use of whole body and extremity monitoring, if required
	• Possess survey instrument calibrated by NRC/Agreement State licensee; or as defined in Appendix I; checked before use
	• 10 CFR 20.1301 surveys
	  – when and where surveys performed
	  – survey records maintained for 3 years				

Item Number and Title	Suggested Response	Applicant's Response			
		Yes	No	Other	
				Yes	No
10 **Radiation Safety Program - Transportation**	The applicant is not required to submit a response to transportation section during the licensing process; this issue will be reviewed during inspection.	Need Not Be Submitted with Application			
10 **Radiation Safety Program - Fixed Gauges Used At Temporary Job Sites**	This is not applicable to the applicant's program. Applicant will not use fixed gauges at temporary job sites. **OR** Procedures will be developed, implemented, maintained and distributed and will meet the Criteria in the section entitled "Radiation Safety Program - Fixed Gauges Used at Temporary Job Sites," in NUREG-155, Vol. 4, dated October 1998.	[ ] N/A [ ]	 [ ]		

Item Number and Title	Suggested Response	Applicant's Response			
		Yes	No	Other	
				Yes	No
10 Radiation Safety Program - Fixed Gauges Used At Temporary Job Sites *(Cont'd)*	**Optional Response** • Develop, implement, maintain, and distribute operating and emergency procedures containing the following elements: – Instructions for transporting radioactive material to ensure compliance with DOT regulations – Instructions for using gauges at temporary job sites and performing routine maintenance according to the manufacturer's or distributor's recommendations and instructions – Instructions for maintaining security during storage and transportation – Instructions to keep gauges under control and immediate surveillance or secured to prevent unauthorized use or access – Steps to take to keep radiation exposures ALARA – Steps to maintain accountability during use – Steps to control access to a potentially damaged gauge – Steps to take, and whom to contact, when a gauge has been lost or damaged. • If gauges are to be installed at temporary job sites, the operating and emergency procedures should contain instructions on the use of personal dosimetry, and survey instruments and conducting surveys. • Provide copies of operating and emergency procedures to all gauge users and maintain copies at each job site.				
10 Radiation Safety Program - Minimization of Contamination	The applicant does not need to provide a response to this item under the following condition. NRC will consider that the above criteria have been met if the applicant's responses meet the criteria for the following sections: Radioactive Material - Sealed Sources and Devices, Facilities and Equipment, Radiation Safety Program - Operating and Emergency Procedures, Radiation Safety Program - Leak Testing, and Waste Management - Gauge Transfer and Disposal.	Need Not Be Submitted with Application			

Item Number and Title	Suggested Response	Applicant's Response			
		Yes	No	Other	
				Yes	No
11 Waste Disposal - Fixed Gauge Disposal & Transfer	The applicant is not required to submit a response to waste management section during the licensing process; however, the licensee should develop, implement, and maintain fixed gauge transfer and disposal procedures in its radiation safety program.	Need Not Be Submitted with Application			

Appendix E

Sample SSD Registration Certificate

Sample SSD Registration Certificate

Appendix F

Typical Duties and Responsibilities of the Radiation Safety Officer

Typical Duties and Responsibilities of the Radiation Safety Officer

The RSO's duties and responsibilities include ensuring radiological safety and compliance with both NRC regulations and the conditions of the license. (See Figure 8.2.) Typically, the RSO's duties and responsibilities include ensuring the following:

- Activities involving licensed material that the RSO considers unsafe are stopped

- Radiation exposures are ALARA

- Development, maintenance, distribution, and implementation of up-to-date operating and emergency procedures

- Individuals that use fixed gauges are properly trained

- Possession, installation, relocation, use, storage, routine maintenance and non-routine operations of fixed gauges are consistent with the limitations in the license, the SSD Registration Certificate(s), manufacturer's or distributor•s recommendations and instructions

- Safety consequences of non-routine operations are analyzed before conducting any such activities that have not been previously analyzed

- Non-routine operations are performed by the manufacturer, distributor or person specifically authorized by the NRC or an Agreement State

- Prospective evaluations are performed demonstrating that unmonitored individuals are not likely to receive, in one year, a radiation dose in excess of 10% of the allowable limits or personnel monitoring devices are provided

- Personnel monitoring devices, if required, are used and exchanged at the proper intervals, and records of the results of such monitoring are maintained

- Documentation is maintained to demonstrate, by measurement or calculation, that the TEDE to the individual member of the public likely to receive the highest dose from the licensed operation does not exceed the annual limit in 10 CFR 20.1301

- Fixed gauges are properly secured

- Notification of proper authorities of incidents such as damage to or malfunction of fixed gauges, fire, loss, or theft

- Investigation of unusual occurrences involving the fixed gauge (e.g., malfunctions or damage), identification of cause(s), implement of appropriate and timely corrective action(s)

- Radiation safety program audits are performed at intervals not to exceed 12 months and development, implement, and documentation of timely corrective actions

- When the licensee identifies violations of regulations or license conditions or program weaknesses, corrective actions are developed, implemented, and documented

- Licensed material is transported according to all applicable DOT requirements

- Licensed material is disposed of properly

- Appropriate records are maintained

- An up-to-date license is maintained and amendment and renewal requests are submitted in a timely manner

- Posting of documents required by 10 CFR 19.11 (Parts 19 and 20, license documents, operating procedures, NRC Form 3 "Notice to Employees"), and 10 CFR 21.6 (Part 21, Section 206 of Energy Reorganization Act of 1974, procedures adopted pursuant to Part 21) or posting a notice indicating where these documents can be examined

Appendix G

Criteria for Acceptable Training for Authorized Users and Radiation Safety Officers

Criteria for Acceptable Training for Authorized Users and Radiation Safety Officers

Course Content

Classroom training may be in the form of lecture, videotape, or self-study emphasizing practical subjects important to safe use of the gauge:

Radiation Safety:

- Radiation vs. contamination

- Internal vs. external exposure

- Biological effects of radiation

- Types and relative hazards of radioactive material possessed

- ALARA concept

- Use of time, distance, and shielding to minimize exposure

- Location of sealed source within the gauge

Regulatory Requirements:

- Applicable regulations

- License conditions, amendments, renewals

- Locations of use and storage of radioactive materials

- Material control and accountability

- Annual audit of radiation safety program

- Transfer and disposal

- Recordkeeping

- Prior events involving fixed gauges

- Handling incidents

- Recognizing and ensuring that radiation warning signs are visible and legible

- Licensing and inspection by regulatory agency

- Need for complete and accurate information

- Employee protection
- Deliberate misconduct

Practical Explanation of the Theory and Operation for Each Gauge Possessed by the Licensee:

- Operating and emergency procedures
- Routine vs. non-Routine maintenance
- Lock-out procedures

On-the-job training must be done under the supervision of an AU or RSO:

- Supervised Hands-on Experience Performing:
    — Operating procedures
    — Test runs of emergency procedures
    — Routine maintenance
    — Lock-out procedures

Training Assessment

Management will ensure that proposed AUs are qualified to work independently with each type of gauge with which they may work. Management will ensure that proposed RSO's are qualified to work independently with and are knowledgeable of the radiation safety aspects of all types of gauges to be possessed by the applicant. This may be demonstrated by written or oral examination or by observation.

Course Instructor Qualifications

Instructor should have:

- Bachelor's degree in a physical or life science or engineering
- Successful completion of a fixed gauge manufacturer's or distributor's course for users (or equivalent)
- Successful completion of an 8 hour radiation safety course; and
- 8 hours hands-on experience with fixed gauges

OR

- Successful completion of a fixed gauge manufacturer's or distributor's course for users (or equivalent)

- Successful completion of 40 hour radiation safety course; and

- 30 hours of hands-on experience with fixed gauges.

OR

- The applicant may submit a description of alternative training and experience for the course instructor.

Note: Additional training is required for those applicants intending to perform non-routine operations such as installation, initial radiation survey, repair, and maintenance of components related to the radiological safety of the gauge, gauge relocation, replacement, and disposal of sealed sources, alignment, or removal of a gauge from service. See Appendix N - "Non-Routine Operations."

Appendix H

Suggested Fixed Gauge Audit Checklist

Suggested Fixed Gauge Audit Checklist

Note: All areas indicated in audit notes may not be applicable to every license and may not need to be addressed during each audit. For example, licensees do not need to address areas which do not apply to their activities and activities which have not occurred since the last audit need not be reviewed at the next audit.

Licensee's name _____   License No. _____

Date of This Audit _____   Date of Last Audit _____

_____   Date _____
(Auditor Signature)

_____   Date _____
(Management Signature)

Audit History

A. Last audit of this location conducted on (date) _____

B. Were previous audits conducted at intervals not to exceed 12 months? [10 CFR 20.1101]

C. Were records of previous audits maintained? [10 CFR 20.2102]

D. Were any deficiencies identified during last two audits or two years, whichever is longer?

E. Were corrective actions taken? (Look for repeated deficiencies).

Organization and Scope of Program

A. If the mailing address or places of use changed, was the license amended?

B. If ownership changed or bankruptcy filed, was NRC prior consent obtained or was NRC notified?

C. Radiation Safety Officer

   1. If the RSO was changed, was license amended?

   2. Does new RSO meet NRC training requirements?

   3. Is RSO fulfilling his/her duties?

   4. To whom does RSO report?

D. If the designated contact person for NRC changed, was NRC notified?

E. Sealed Sources and Devices

    1. Does the license authorize all of the NRC regulated radionuclides contained in gauges?

    2. Are the gauges as described in the Sealed Source and Device (SSD) Registration Certificate?

    3. Have copies of (or access to) SSD Certificates?

    4. Have manufacturers' or distributor•s manuals for operation and maintenance? [10 CFR 32.210]

    5. Are the actual uses of gauges consistent with the authorized uses listed on the license?

    6. Are the location of the gauges compatible with the "Conditions of Normal Use" and "Limitations and/or Other Considerations of Use" on the SSD Registration Certificates?

Training and Instructions to Workers

A. Were all workers who are likely to exceed 1 mSv (100 mrem) in a year instructed per [10 CFR 19.12]? Refresher training provided, as needed [10 CFR 19.12]? Records maintained?

B. Did each AU receive training and instruction given at the time of gauge installation or equivalent training and instruction before using gauges?

C. Are training records maintained for each AU?

D. Did individuals who perform non-routine operations receive training before performing these operations?

E. Did interviews with AUs reveal that they know the emergency procedures?

F. Did this audit include observations of AUs using the gauge?

G. Did this audit include observations of workers performing routine cleaning and lubrication on the gauge?

H. HAZMAT training provided, if required? [49 CFR 172.700, 172.701, 172.702, 172.703, 172.704]

Radiation Survey Instruments

A. If the licensee is required to possess a survey meter, does it meet the NRC's criteria? [10 CFR 20.1501]

B. Are calibration records maintained [10 CFR 20.2103(a)]?

Gauge Inventory

A. Is a record kept showing the receipt of each gauge? [10 CFR 30.51(a)(1)]

B. Are all gauges physically inventoried every six months?

C. Are records of inventory results with appropriate information maintained?

Personnel Radiation Protection

A. Are ALARA considerations incorporated into the radiation protection program? [10 CFR 20.1101(b)]

B. Were prospective evaluations performed showing that unmonitored individuals receive ≤10% of limit? [10 CFR 20.1502(a)]

C. Did unmonitored individuals' activities change during the year which could put them over 10% of limit?

D. If yes to C. above, was a new evaluation performed?

E. Is external dosimetry required (individuals likely to receive >10% of limit,)? And is dosimetry provided to these individuals?

   1. Is the dosimetry supplier NVLAP approved? [10 CFR 20.1501(c)]

   2. Are the dosimeters exchanged monthly for film badges and quarterly for TLD's?

   3. Are dosimetry reports reviewed by the RSO when they are received?

   4. Are the records NRC Forms or equivalent? [10 CFR 20.2104(d), 20.2106(c)]

       a. NRC-Form 4 "Cumulative Occupational Exposure History" completed?

       b. NRC-Form 5 "Occupational Exposure Record for a Monitoring Period" completed?

   5. Declared pregnant worker/embryo/fetus

       a. If a worker declared her pregnancy, did licensee comply with [10 CFR 20.1208]?

       b. Were records kept of embryo/fetus dose per [10 CFR 20.2106(e)]?

F. Are records of exposures, surveys, monitoring, and evaluations maintained [10 CFR 20.2102, 20.2103, 20.2106]

Public Dose

A. Is public access to gauges controlled in a manner to keep doses below 1 mSv (100 mrem) in a year? 10 CFR 20.1301(a)(1)]

B. Has a survey or evaluation been performed per 10 CFR 20.1501(a)? Have there been any additions or changes to the storage, security, or use of surrounding areas that would necessitate a new survey or evaluation?

C. Do unrestricted area radiation levels exceed 0.02 mSv (2 mrem) in any one hour? [10 CFR 20.1301(a)(2)]

D. Is gauge access controlled in a manner that would prevent unauthorized use or removal? [10 CFR 20.1801]

E. Records maintained? [10 CFR 20.2103, 20.2107]

Operating and Emergency Procedures

A. Have operating and emergency procedures been developed?

B. Do they contain the required elements?

C. Does each individual working with the gauges have a current copy of the operating and emergency procedures (including lock-out procedures and emergency telephone numbers)?

D. Is a lock-out warning sign posted at each entryway to an area where it is possible to be exposed to the beam?

E. Did any emergencies occur?

  1. If so, were they handled properly?

  2. Were appropriate corrective actions taken?

  3. Was NRC notification or reporting required? [10 CFR 20.2201, 20.2202, 20.2203]

Leak Tests

A. Was each sealed source leak tested every 6 months or at other prescribed intervals?

B. Was the leak test performed according to the license?

C. Are records of results retained with the appropriate information included?

D. Were any sources found leaking and if yes, was NRC notified?

Maintenance of Gauges

A. Are manufacturer's or distributor•s procedures followed for routine cleaning and lubrication of gauge?

B. Was each on-off mechanism tested for proper operation every 6 months or at other prescribed intervals?

C. Are repair and maintenance of components related to the radiological safety of the gauge performed by the manufacturer, distributor or person specifically authorized by the NRC or an Agreement State and according to license requirements (e.g., extent of work, procedures, dosimetry, survey instrument, compliance with 10 CFR 20.1301 limits)?

D. Are labels, signs, and postings identifying gauges containing radioactive material, radiation areas, and lock-out procedures/warnings clean and legible?

Transportation

(*Note:* This section will not apply if you have not transported gauges during the period covered by this audit.)

A. DOT-7A or other authorized packages used? [49 CFR 173.415, 173.416(b)]

B. Package performance test records on file?

C. Special form sources documentation? [49 CFR 173.476(a)]

D. Package has two labels (ex. Yellow-II) with TI, Nuclide, Activity, and Hazard Class? [49 CFR 172.403, 173.441]

E. Package properly marked? [49 CFR 172.301, 172.304, 172.310, 172.324]

F. Package closed and sealed during transport? [49 CFR 173.475(f)]

G. Shipping papers prepared and used? [49 CFR 172.200(a)]

H. Shipping papers contain proper entries? {Shipping name, Hazard Class, Identification Number (UN Number), Total Quantity, Package Type, Nuclide, RQ, Radioactive Material, Physical and Chemical Form, Activity (SI units required), category of label, TI, Shipper's Name, Certification and Signature, Emergency Response Phone Number, Cargo Aircraft Only (if applicable)} [49 CFR 172.200, 172.201, 172.202, 172.203, 172.204, 172.604]

I. Shipping papers within drivers reach and readily accessible during transport? [49 CFR 177.817(e)].

J. Package secured against movement? [49 CFR 177.834]

K. Placards on vehicle, if needed? [49 CFR 172.504]

L. Proper overpacks, if used? [49 CFR 173.25]

M. Any incidents reported to DOT? [49 CFR 171.15, 171.16]

Auditor's Independent Survey Measurements (If Made)

A. Describe the type, location, and results of measurements. Does any radiation level exceed regulatory limits? [10 CFR 20.1501(a) & 20.1502(a)]

Notification and Reports

A. Was any radioactive material lost or stolen? Were reports made? [10 CFR 20.2201, 30.50]

B. Did any reportable incidents occur? Were reports made? [10 CFR 20.2202, 21.21, 30.34, 30.36, 30.50]

C. Did any overexposures and high radiation levels occur? Reported? [10 CFR 20.2203, 30.50]

D. If any events (as described in items a through c above) did occur, what was root cause? Were corrective actions appropriate?

E. Is the management/RSO/shift foreman licensee aware of telephone number for NRC Emergency Operations Center? [(301) 816-5100]

Posting and Labeling

A. NRC-Form 3 "Notice to Workers" posted? [10 CFR 19.11]

B. NRC regulations, license documents posted or a notice posted? [10 CFR 19.11, 21.6]

C. Other posting and labeling? [10 CFR 20.1902, 20.1904]

Record Keeping for Decommissioning

A. Records kept of information important to decommissioning? [10 CFR 30.35(g)]

B. Records include all information outlined in [10 CFR 30.35(g)]?

Bulletins and Information Notices

A. NRC Bulletins, NRC Information Notices, NMSS Newsletters, received?

B. Appropriate training and action taken in response?

Special License Conditions or Issues

A. Did auditor review special license conditions or other issues (e.g., non-routine operations)?

Deficiencies Identified in Audit; Corrective Actions

A. Summarize problems/deficiencies identified during audit.

B. If problems/deficiencies identified in this audit, describe corrective actions planned or taken. Are corrective actions planned or taken at ALL licensed locations (not just location audited)? Include date(s) when corrective actions are implemented.

C. Provide any other recommendations for improvement.

Evaluation of Other Factors

A. Senior licensee management is appropriately involved with the radiation protection program and/or RSO oversight?

B. RSO has sufficient time to perform his/her radiation safety duties?

C. Licensee has sufficient staff to support the radiation protection program?

Appendix I

Model Survey Instrument Calibration Program

Model Survey Instrument Calibration Program

Training

Before independently calibrating survey instruments, an individual should complete both classroom and on-the-job training as follows:

- Classroom training may be in the form of lecture, videotape, or self-study and will cover the following subject areas:

  — Principles and practices of radiation protection

  — Radioactivity measurements, monitoring techniques, and the use of instruments

  — Mathematics and calculations basic to using and measuring radioactivity

  — Biological effects of radiation.

- On-the-job training will be considered complete if the individual has:

  — Observed authorized personnel performing survey instrument calibration; and

  — Conducted survey meter calibrations under the supervision, and in the physical presence of an individual already authorized to perform calibrations.

Facilities and Equipment

- To reduce doses received by individuals not calibrating instruments, calibrations will be conducted in an isolated area of the facility or at times when no one else is present

- Individuals conducting calibrations will wear assigned dosimetry

- Individuals conducting calibrations will use a calibrated and operable survey instrument to ensure that unexpected changes in exposure rates are identified and corrected.

Model Procedure for Calibrating Survey Instruments

- A radioactive sealed source(s) will be used for calibrating survey instruments, and this source will:

  — Approximate a point source

  — Have its apparent source activity or the exposure rate at a given distance traceable by documented measurements to a standard certified to be within ± 5% accuracy by National Institutes of Standards and Technology (NIST)

— Contain a radionuclide which emits radiation of identical or similar type and energy as the sealed sources that the instrument will measure

— Be strong enough to emit a radiation field that is representative of the field being emitted by the gauge. For calibration of instruments intended to measure gamma radiation, the exposure rate should be at least 30 mR/hour (7.7 microcoulomb/kilogram per hour) at 100 cm [e.g., 3.1 gigabecquerels (85 millicuries) of Cs-137 or 780 megabecquerels (21 millicuries) of Co-60].

- Inverse square and radioactive decay laws must be used to correct changes in exposure rate due to changes in distance or source decay.

- A record must be made of each survey meter calibration.

- A single point on a survey meter scale may be considered satisfactorily calibrated if the indicated exposure rate differs from the calculated exposure rate by less than ±20%.

- There are three kinds of scales frequently used on radiation survey meters. They are calibrated either as described in ANSI N323A-1996, "American National Standard Radiation Protection Instrumentation Test and Calibration - Portable Survey Instruments," or as follows:

— Meters on which the user selects a linear scale must be calibrated at not fewer that two points on each scale. The points will be at approximately 1/3 and 2/3 of the decade.

— Meters that have a multidecade logarithmic scale must be calibrated at one point (at the least) on each decade and not fewer than two points on one of the decades. Those points will be approximately 1/3 and 2/3 of the decade.

— Meters that have an automatically ranging digital display device for indicating exposure rates must be calibrated at one point (at the least) on each decade and at no fewer than two points on one of the decades. Those points should be at approximately 1/3 and 2/3 of the decade.

- Readings above 200 mR/hour (50 microcoulomb/kilogram per hour) need not be calibrated. However, higher scales should be checked for operation and approximately correct response.

- Survey meter calibration reports will indicate the procedure used and the results of the calibration. The reports will include:

— The owner or user of the instrument

— A description of the instrument that includes the manufacturer's name, model number, serial number, and type of detector

— A description of the calibration source, including the exposure rate at a specified distance on a specified date, and the calibration procedure

— For each calibration point, the calculated exposure rate, the indicated exposure rate, the deduced correction factor (the calculated exposure rate divided by the indicated exposure rate), and the scale selected on the instrument

— The exposure reading indicated with the instrument in the "battery check" mode (if available on the instrument)

— For instruments with external detectors, the angle between the radiation flux field and the detector (i.e., parallel or perpendicular)

— For instruments with internal detectors, the angle between radiation flux field and a specified surface of the instrument

— For detectors with removable shielding, an indication whether the shielding was in place or removed during the calibration procedure

— The exposure rate from a check source, if used

— The signature of the individual who performed the calibration and the date on which the calibration was performed.

• The following information will be attached to the instrument as a calibration sticker or tag:

— The source that was used to calibrate the instrument

— The proper deflection in the battery check mode (unless this is clearly indicated on the instrument)

— For each scale or decade not calibrated, an indication that the scale or decade was checked only for function but not calibrated

— The date of calibration and the next calibration due date

— The apparent exposure rate from the check source, if used.

References: Detailed information about survey instrument calibration may be obtained by referring to ANSI N323A-1996, "American National Standard Radiation Protection Instrumentation Test and Calibration -Portable Survey Instruments." Copies may be ordered electronically at <http://www.ansi.org> or by writing to ANSI, 1430 Broadway, New York, NY 10018.

See the Notice of Availability (on the inside front cover of this report) to obtain copies of Draft RG FC 413-4, "Guide for the Preparation of Applications for Licenses for the Use of Radioactive Materials in Calibrating Radiation Survey and Monitoring Instruments," dated June 1985.

Appendix J

Guidance for Demonstrating That Unmonitored Individuals Are Not Likely to Exceed 10 Percent of the Allowable Limits

Guidance for Demonstrating That Unmonitored Individuals Are Not Likely to Exceed 10 Percent of the Allowable Limits

Dosimetry is required for individuals likely to receive, in 1 year from sources external to the body, a dose in excess of 10% of the applicable regulatory limits in 10 CFR 20.1201. To demonstrate that dosimetry is *not* required, a licensee needs to perform a prospective evaluation to demonstrate that its workers are not likely to exceed 10% of the applicable annual limits.

The most common way that individuals might exceed 10% of the applicable limits is by performing frequent routine maintenance on the gauge. However, for most gauges even these activities result in the individual receiving minimal doses. Before allowing workers to perform these tasks, a licensee will need to evaluate the doses which its workers might receive to assess whether dosimetry is required; this is a prospective evaluation.

Example

One gauge manufacturer has estimated the doses to the extremities and whole body of a person replacing the assay plate on one of its series of gauges. Each gauge in the series is authorized to contain up to 7.4 gigabecquerels (200 millicuries) of Cs-137. The manufacturer based its estimate on observations of individuals performing the recommended procedure according to good radiation safety practices. The manufacturer had the following types of information:

- Time needed to perform the entire procedure (e.g., 15 minutes)

- Expected dose rate received by the whole body of the individual, associated with the shielded source and determined using measured or manufacturer-determined data (e.g., 0.02 mSv/hr [2 mrem/hr] at 46 cm [18.1 in] from the shield)

- Time the hands were exposed to the shielded source (e.g., 6 min)

- Expected dose rate received by the extremities of the individual, associated with the shielded source and determined using measured or manufacturer-determined data on contact with the shield (e.g., 0.15 mSv/hr [15 mrem/hr])

From this information, the manufacturer estimated that the individual performing each routine cleaning and lubrication could receive the following:

- Less than 0.005 mSv (0.5 mrem) TEDE (whole body) and

- 0.015 mSv (1.5 mrem) to the hands.

The applicable TEDE (whole body) limit is 50 mSv (5 rems) per year and 10% of that value is 5 mSv (500 millirems) per year. If one of these procedures delivers 0.005 mSv (0.5 mrem), then an individual could perform 1,000 of these procedures each year and remain within 10% of the applicable limit.

The applicable shallow-dose equivalent (SDE) (extremities) is 500 mSv (50 rems) is 500 mSv (50 rems) per year and 10% of that value is 50 mSv (5 rems or 5000 millirems) per year. If one of these procedures delivers 0.015 mSv (1.5 mrem), then an individual could perform 3,333 of these procedures each year and remain within 10% of the applicable limit.

Based on the above specific situation, no dosimetry is required if a worker performs fewer than 1,000 routine maintenance procedures per year.

Guidance to Licensees

Licensees who wish to demonstrate that they are not required to provide dosimetry to their workers need to perform prospective evaluations similar to that shown in the example above. The expected dose rates, times, and distances used in the above example may not be appropriate to individual licensee situations. In their evaluations, licensees need to use information appropriate to the type(s) of gauge(s) they intend to use; this information is generally available from the gauge manufacturer or the SSD Registration Certificate maintained by the NRC and Agreement States.

Table J.1 may be helpful in performing a prospective evaluation.

Licensees should review evaluations periodically and revise them as needed. Licensees need to check assumptions used in their evaluations to ensure that they continue to be up-to-date and accurate. For example, if workers become lax in following good radiation safety practices, perform the task more slowly than estimated, work with new gauges containing sources of different activities or radionuclides, or use modified procedures, the licensee would need to conduct a new evaluation.

Table J.1   Dosimetry Evaluation

Dosimetry Evaluation for _____	Model _____	Gauge _____	
A.	Time needed to perform the entire routine maintenance procedure.	_____ minutes/60	_____ hour
B.	Expected whole body dose rate received by the individual, determined using exposure rates measured on contact with the gauge while the sealed source is in the shielded position.		_____ mrem/hr
C.	Time the <u>hands</u> were exposed to the shielded source.	_____ minutes/60	_____ hour
D.	Expected extremity dose rate received by the individual, determined using measured or manufacturer-provided data for the shielded source at the typical distance from the hands to the shielded source.		_____ mrem/hr

Formula: (_____ # hours in Row A) x (_____ mrem/hr in Row B) = (_____ mrem per routine procedure) x (_____ # of routine maintenance procedures each year) = _____ mrem *Whole Body Dose

Formula: (_____ # hours in Row C) x (_____ mrem/hr in Row D) = (_____ mrem per routine procedure) x (_____ # of routine maintenance procedures each year) = _____ mrem **Extremity Dose

*   Expected Whole Body Doses *less than* 500 mrem requires no dosimetry
** Expected Extremity Doses *less than* 5000 mrem requires no dosimetry

Appendix K

Guidance for Demonstrating That Individual Members of the Public Will Not Receive Doses Exceeding the Allowable Limits

Guidance for Demonstrating That Individual Members of the Public Will Not Receive Doses Exceeding the Allowable Limits

Licensees must ensure that:

- The radiation dose received by individual members of the public does not exceed 1 millisievert (1 mSv) [100 millirems (100 mrem)] in one calendar year resulting from the licensee's possession and/or use of licensed materials.

> Members of the public include persons who live, work, or may be near locations where fixed gauges are used or stored and employees whose assigned duties do not include the use of licensed materials and who work in the vicinity where gauges are used or stored.

The radiation dose in unrestricted areas does not exceed 0.02 mSv (2 mrem) in any one hour.

> Typical unrestricted areas may include offices, shops, laboratories, a nearby walkway, an area near the gauge that requires frequent maintenance, areas outside buildings, and nonradioactive equipment storage areas. The licensee does not control access to these areas for purposes of controlling exposure to radiation or radioactive materials. However, the licensee may control access to these areas for other reasons such as security.

Licensees must show compliance with both portions of the regulation. Calculations or a combination of calculations and measurements (e.g., using an environmental TLD) are often used to prove compliance.

Calculational Method

For ease of use by most fixed gauge licensees, the examples in this Appendix use conventional units. The conversions to SI units are as follows: 1 ft = 0.305 m; 1 mrem = 0.01 mSv.

The calculational method takes a tiered approach, going through a three-part process starting with a worst case situation and moving toward more realistic situations. It makes the following simplifications:

- each gauge is a point source;

- typical radiation levels encountered when the source is in the shielded position are taken from either the Sealed Source & Device (SSD) Registration Certificate or the manufacturer's literature; and

- no credit is taken for any shielding found between the gauges and the unrestricted areas.

Part 1 of the calculational method is simple but conservative. It assumes that an affected member of the public is present 24 hours a day and uses only the inverse square law to determine if the distance between the gauge and the affected member of the public is sufficient to show compliance with the public dose limits. Part 2 considers not only distance, but also the time that the affected member of the public is actually in the area under consideration. Part 3 considers distance and the portion of time that both the gauge and the affected member of the public are present. Using this approach, licensees make only those calculations that are needed to demonstrate compliance. In many cases licensees will need to use the calculational method through Part 1 or Part 2. The results of these calculations typically result in higher radiation levels than would exist at typical facilities, but provide a method for estimating conservative doses which could be received.

Example 1

To better understand the calculational method, we will look at ABC Bottling, Inc., a fixed gauge licensee. Yesterday, while on a walk-through during product changeover, the company's president noted that three new gauges will be very close to a bottling control panel where a quality control supervisor, a worker who does not work with fixed gauges, works. The company's president asked Joe, the Radiation Safety Officer (RSO), to determine if the company is complying with NRC's regulations.

Joe measures the distances from each gauge to the bottling control panel and looks up in the manufacturer's literature the radiation levels individuals would encounter for each gauge. Figure K.1 is Joe's sketch of the areas in question, and Table K-1 summarizes the information Joe has on each gauge.

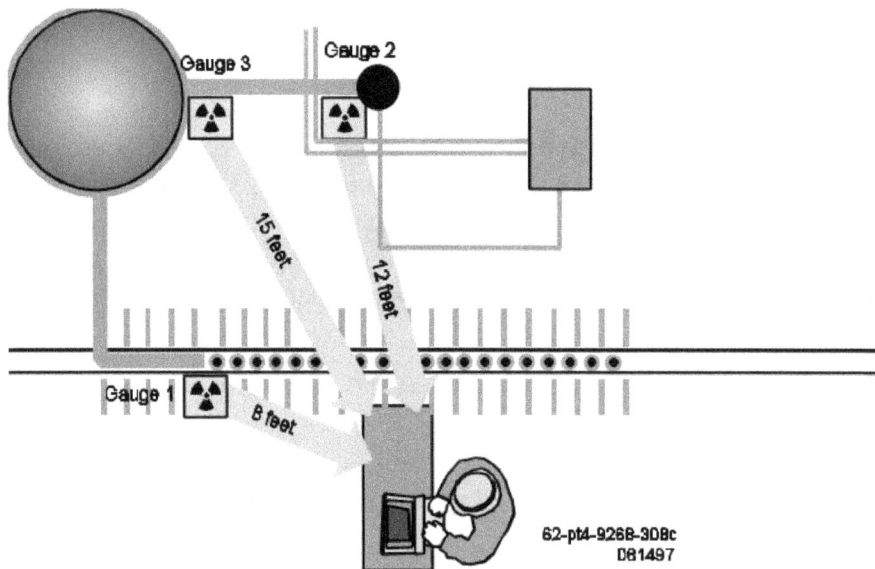

Figure K.1 Diagram of Bottling Line and Fixed Gauges. *This sketch shows the areas described in Examples 1 and 2.*

Table K.1 Information Known about Each Gauge

Description of Known Information	Gauge 1	Gauge 2	Gauge 3
Where gauge is located	Gauge on bottling line	Gauge on main feed line	Gauge on tank
Dose rate in mrem/hr encountered at specified distance from the gauge (from manufacturers literature)	2 mrem/hr at 1 ft	8 mrem/hr at 1 ft	2 mrem/hr at 3 ft
Distance in ft to bottling control panel	8 ft	12 ft	15 ft

Example 1: Part 1

Joe's first thought is that the distance between the gauges and the bottling control panel may be sufficient to show compliance with the regulation in 10 CFR 20.1301. So, taking a worst case approach, he assumes: 1) the gauges are constantly present (i.e., 24 hr/d), 2) all three gauges are on (i.e., shutters are open), and 3) a quality control (QC) supervisor, a worker who does not work with the fixed gauges, is constantly sitting at the control panel (i.e., 24 hr/d). Joe proceeds to calculate the dose the QC supervisor might receive hourly and yearly from each gauge as shown in Tables K-2, K-3, and K-4 below.

Table K.2 Calculational Method, Part 1: Hourly and Annual Dose Received from Gauge 1

Step No.	Description	Gauge 1	
		Input Data	Results
1	Dose received in an hour at known distance from gauge (e.g., from manufacturer's data), in mrem/hr	2	2
2	Square of the distance (ft) at which the Step 1 rate was measured, in ft^2	$(1)^2$	1
3	Square of the distance (ft) from the gauge to the bottling control panel in an unrestricted area, in ft^2	$(8)^2$	64
4	Multiply the results of Step 1 by the results of Step 2 (this is an intermediate result)	2 x 1 =2	

		Gauge 1	
Step No.	**Description**	**Input Data**	**Results**
5	Divide the result of Step 4 by the result of Step 3 to calculate the dose received by the worker at the bottling control panel, **HOURLY DOSE RECEIVED FROM GAUGE 1**, in mrem in an hour.	2/64 = **0.031**	
6	Multiply the result of Step 5 by 24 hr/d x 365 d/yr = **MAXIMUM ANNUAL DOSE RECEIVED FROM GAUGE 1**, in mrem in a year.	0.031 x 24 x 365 = 0.031 x 8760 = **272**	

Table K.3   Calculational Method, Part 1: Hourly and Annual Dose Received from Gauge 2

		Gauge 2	
Step No.	**Description**	**Input Data**	**Results**
1	Dose received in an hour at known distance from gauge (e.g., from manufacturer's data), in mrem/hr	8	8
2	Square of the distance (ft) at which the Step 1 rate was measured, in ft^2	$(1)^2$	1
3	Square of the distance (ft) from the gauge to the bottling control panel in an unrestricted area, in ft^2	$(12)^2$	144
4	Multiply the results of Step 1 by the results of Step 2 (this is an intermediate result)	8 x 1 = **8**	
5	Divide the result of Step 4 by the result of Step 3 to calculate dose received in an hour by the worker at the bottling control panel, **HOURLY DOSE RECEIVED FROM GAUGE 2**, in mrem in an hour	8/144 = **.056**	
6	Multiply the result of Step 5 by 24 hr/d x 365 d/yr = **MAXIMUM ANNUAL DOSE RECEIVED FROM GAUGE 2**, in mrem in a year	0.056 x 24 x 365 = 0.056 x 8760 = **491**	

Table K.4 Calculational Method, Part 1: Hourly and Annual Dose Received from Gauge 3

Step No.	Description	Input Data	Results
		Gauge 3	
1	Dose received in an hour at known distance from gauge (e.g., from manufacturer's data), in mrem/hr	2	2
2	Square of the distance (ft) at which the Step 1 rate was measured, in ft^2	$(3)^2$	9
3	Square of the distance (ft) from the gauge to bottling control panel in an unrestricted area, in ft^2	$(15)^2$	225
4	Multiply the results of Step 1 by the results of Step 2 (this is an intermediate result)	2 x 9 =18	
5	Divide the result of Step 4 by the result of Step 3 to calculate dose received by the worker at the bottling control panel, **HOURLY DOSE RECEIVED FROM GAUGE 3**, in mrem in an hour	18/225 = **0.08**	
6	Multiply the result of Step 5 by 24 hr/d x 365 d/yr = **MAXIMUM ANNUAL DOSE RECEIVED FROM GAUGE 3**, in mrem in a year	0.08 x 24 x 365 = 0.08 x 8760 = **701**	

To determine the total hourly and total annual dose received, Joe adds the pertinent data from the preceding tables.

Table K.5 Calculational Method, Part 1: Total Hourly and Annual Dose Received from Gauges 1, 2, and 3

Step No.	Description	Gauge 1	Gauge 2	Gauge 3	Sum
7	**TOTAL HOURLY DOSE RECEIVED** from Step 5 of Tables K-2, K-3, and K-4, in mrem in an hour	0.031	0.056	0.08	0.031 + 0.056 + 0.08 = **0.167**
8	**TOTAL ANNUAL DOSE RECEIVED** from Step 6 of Tables K-2, K-3, and K-4, in mrem in a year	272	491	701	272 + 491 + 701 = **1464**

Note: The Sum in Step 7 demonstrates compliance with the 2 mrem in any one hour limit. Reevaluate if assumptions change. If the Sum in Step 8 exceeds 100 mrem/yr, proceed to Part 2 of the calculational method.

At this point, Joe is pleased to see that the total dose that an individual could receive in any one hour is only 0.167 mrem, but notes that an individual could receive a dose of 1,464 mrem in a year, much higher than the 100 mrem limit.

Example 1: Part 2

Joe reviews his assumptions and recognizes that the QC supervisor is not at the bottling control panel 24 hr/d. He decides to make a realistic estimate of the number of hours the QC supervisor would be present at the bottling control panel, keeping his other assumptions constant (i.e., the gauges are constantly present (i.e., 24 hr/d), all three gauges remain on (i.e., shutter is open). He then recalculates the annual dose received.

Table K.6   Calculational Method, Part 2: Annual Dose Received from Gauges 1, 2, and 3

Step No.	Description	Results
9	A. Average number of hours per day that individual spends in area of concern (e.g., worker present at bottling control panel 5 hr/day; the remainder of the day the worker is away from the area performing other duties that are not in the vicinity of gauges)	5
	B. Average number of days per week in area (e.g., worker is part time and works 3 days/week)	3
	C. Average number of weeks per year in area (e.g., worker works all year)	52
10	Multiply the results of Step 9.A. by the results of Step 9.B. by the results of Step 9.C. = **AVERAGE NUMBER OF HOURS IN AREA OF CONCERN PER YEAR**	5 x 3 x 52 = **780**
11	Multiply the sum in Step 7 by the results of Step 10 = **ANNUAL DOSE RECEIVED FROM GAUGES CONSIDERING REALISTIC ESTIMATE OF TIME SPENT IN AREA OF CONCERN**, in mrem in a year	0.167 x 780 = **130**

Note: If Step 11 exceeds 100 mrem in a year, proceed to Part 3 of the calculational method.

Although Joe is pleased to note that the calculated annual dose received is significantly lower, he realizes it still exceeds the 100 mrem in a year limit.

Example 1, Part 3

Again Joe reviews his assumptions and recognizes that Gauge 3 will only be used on the process line during product changeovers and Gauge 2 has different radiation levels depending on whether the gauge is in the on or off position (i.e., shutter is open or closed). As he examines the situation, he realizes he must consider each gauge individually.

Table K.7   Calculational Method, Part 3: Summary of Information

INFORMATION ON GAUGES:

- **Gauge 1:** operates continuously (24 hrs/day) on the bottling line.
- **Gauge 2:** operates (in the "on" position) while the tank is being filled, approximately 1 hour during the time the worker is present. When the pipe is not filling the tank, the gauge is in the "off" position. While in the "off" position, the radiation level around the gauge drops to 2 mrem/hr at 1ft, one-fourth of the radiation level as when the gauge is in the "on" position.
- **Gauge 3:** is only used on the process line during product changeovers, 4 weeks per year. While affixed, it operates continuously (24 hrs/day).

INFORMATION FROM EXAMPLE 1, PART 2, ON WHEN THE WORKER IS PRESENT AT THE BOTTLING CONTROL PANEL:

- 5 hours per day
- 3 days per week
- 52 weeks per year

Table K.8   Calculational Method, Part 3: Annual Dose Received from Gauges 1, 2, and 3

Step No.	Description	Gauge 1	Gauge 2 "On"	Gauge 2 "Off"	Gauge 3
12	Average number of **hours per day** gauge operates when worker is present at the bottling control panel	5	1	4	5
13	Average number of **days per week** gauge operates when worker is present at the bottling control panel	3	3	3	3
14	Average number of **weeks per year** gauge operates when worker is present at the bottling control panel	52	52	52	4

Step No.	Description	Gauge 1	Gauge 2 "On"	Gauge 2 "Off"	Gauge 3
15	Multiply the results of Step 12 by the results of Step 13 by the results of Step 14 = **TOTAL HOURS EACH GAUGE OPERATED PER YEAR WHILE WORKER IS PRESENT AT BOTTLING CONTROL PANEL**	5x3x52 = **780**	1x3x52 = **156**	4x3x52 = **624**	5x3x4 = **60**
16	Multiply the results of Step 15 by the results of Step 7 (for Gauge 2 in the "off" position, the radiation level drops to 1/4th, so divide the results of Step 7 by 4) = **ANNUAL DOSE RECEIVED FROM EACH GAUGE, in mrem in a year**	780 x 0.031 = **24**	156 x 0.056 = **8.7**	624 x (0.056/4) = **8.7**	60 x 0.08 = **4.8 in mrem in a year**
17	Sum the results of Step 16 for each gauge = **TOTAL ANNUAL DOSE RECEIVED CONSIDERING REALISTIC ESTIMATE OF TIME SPENT IN AREA OF CONCERN AND TIME GAUGE OPERATES, in mrem in a year**	24 + 8.7 + 8.7 + 4.8 = **46.2**			

Note: If the result in Step 17 is greater than 100 mrem/yr, the licensee must take corrective actions.

Joe is pleased that the result in Step 17 shows compliance with the 100 mrem/yr limit. Had the result in Step 17 been higher than 100 mrem/yr, then Joe could have done one or more of the following:

- Consider whether the assumptions used to determine occupancy and the time each gauge operates are accurate, revise the assumptions as needed, and recalculate using the new assumptions

- Calculate the effect of any shielding located between the gauges and the bottling control panel — such calculation is beyond the scope of this Appendix

- Take corrective action (e.g., add shielding, move the bottling control panel) and perform new calculations to demonstrate compliance

- Train the QC supervisor as required by 10 CFR 19.12.

Note that in the example, Joe evaluated the unrestricted area at the bottling control panel. Licensees also need to make similar evaluations for other unrestricted areas and to keep in mind the ALARA principle, taking reasonable steps to keep radiation dose received below regulatory requirements. In addition, licensees need to be alert to changes in situations (e.g., adding a gauge to the process line, changing the QC supervisor's schedule, or changing the estimate of the portion of time spent at the bottling control panel) and to perform additional evaluations, as needed.

> RECORD KEEPING: 10 CFR 20.2107 requires licensees to maintain records demonstrating compliance with the dose limits for individual members of the public.

Combination Measurement - Calculational Method

This method, which allows the licensee to take credit for shielding between the gauge and the area in question, begins by measuring radiation levels in the areas, as opposed to using manufacturer-supplied rates at a specified distance from each gauge. These measurements must be made with calibrated survey meters sufficiently sensitive to measure background levels of radiation. A maximum dose of 1 mSv (100 mrem) received by an individual over a period of 2080 hours (i.e., a work year of 40 hr/wk for 52 wk/yr) is equal to less than 0.5 microsievert (0.05 mrem) per hour.

> This rate is well below the minimum sensitivity of most commonly available G-M survey instruments.

Instruments used to make measurements for calculations must be sufficiently sensitive. An instrument equipped with a scintillation-type detector (e.g., NaI(Tl)) or a micro-R meter used in making very low gamma radiation measurements should be adequate.

Licensees may also choose to use environmental TLDs. TLDs used for personnel monitoring (e.g., LiF) may not have sufficient sensitivity for this purpose. Generally, the minimum reportable dose received is 0.1 mSv (10 mrem). Suppose a TLD monitors dose received and is changed once a month. If the measurements are at the minimum reportable level, the annual dose received could have been about 1.2 mSv (120 mrem), a value in excess of the 1 mSv/yr (100 mrem/yr) limit. If licensees use TLDs to evaluate compliance with the public dose limits, they should consult with their TLD supplier and choose more sensitive TLDs, such as those containing CaF2 that are used for environmental monitoring. This direct measurement method would provide a definitive measurement of actual radiation levels in unrestricted areas without any restrictive assumptions. Records of these measurements can then be evaluated to ensure that rates in unrestricted areas do not exceed the 1 mSv/yr (100 mrem/yr) limit.

Example 2

As in Example 1, Joe is the RSO for ABC Bottling, Inc., a fixed gauge licensee. The company has three gauges located near a bottling control panel which is operated by a worker who does not work with the fixed gauges. See Figure K.1 and Table K-1 for information. Joe wants to see if the company complies with the public dose limits at the bottling control panel.

Joe placed an environmental TLD badge at the bottling control panel for 30 days. The TLD processor sent Joe a report indicating the TLD received 100 mrem.

Table K.9 Combination Measurement - Calculational Method

Step No.	Description	Input Data and Results
Part 1		
1	**Dose** received by TLD, in mrem	**100**
2	Total hours TLD exposed	24 hr/d x 30 d/mo = **720**
3	Divide the results of Step 1 by the results of Step 2 to determine **HOURLY DOSE RECEIVED**, in mrem in an hour	**0.14**
4	Multiply the results of Step 3 by 365 d/yr x 24 hr/d = 8760 hours in one year = **MAXIMUM ANNUAL DOSE RECEIVED FROM GAUGES**, in mrem in a year	365 x 24 x 0.14 = 8760 x 0.14 = **1226**

Note: For the conditions described above, Step 3 indicates that the dose received in any one hour is less than the 2 mrem in any one hour limit. However, if there are any changes, then the licensee would need to reevaluate the potential doses which could be received in any one hour. Step 4 indicates that the annual dose received would be much greater than the 100 mrem in a year allowed by the regulations.

Part 2

At this point Joe can adjust for a realistic estimate of the time the worker spends at the bottling control panel as he did in Part 2 of Example 1.

Part 3

If the results of Joe's evaluation in Part 2 show that the annual dose received in a year exceeds 100 mrem, then he can make adjustments for realistic estimates of the time spent in the area of concern as in Part 3 of Example 1. (Recall that the TLD measurement was made while all the gauges were operating; i.e., 24 hr/d for the 30 days that the TLD was in place.)

Appendix L

Operating and Emergency Procedures

Operating and Emergency Procedures

Operating Procedures:

- If personnel dosimetry is provided:
    - Always wear your assigned thermoluminescent dosimeter (TLD) or film badge when using the gauge.
    - Never wear another person's TLD or film badge.
    - Never store your TLD or film badge near the gauge.

- Use the gauge according to the manufacturer's or distributor•s instructions and recommendations. Perform routine cleaning and maintenance according to the manufacturer's or distributor•s instructions and recommendations.
- Test each gauge for the proper operation of the on-off mechanism (shutter) and indicator, if any, at intervals not to exceed 6 months or as specified in the SSD certificate.
- Do not touch the unshielded source with your fingers, hands, or any part of your body.
- Do not place hands, fingers, feet, or other body parts in the radiation field from an unshielded source.
- Post a radiation warning sign at each entryway to an area where it is possible to be exposed to the beam.
- Prevent employees from entering the radiation beam during maintenance, repairs, or work in, on, or around the bin, tank, or hopper on which the device is mounted by developing lock-out procedures. These procedures should specify who will be responsible for ensuring that the lock-out procedures are followed.
- Prevent unauthorized access, removal, or use of the gauge.
- After making changes affecting the gauge (e.g., changing the location of gauges , removing shielding, adding gauges, changing the occupancy of adjacent areas,), reevaluate compliance with public dose limits and ensure proper security of gauges.
- Conduct a physical inventory every 6 months to account for all sealed sources and devices.

Emergency Procedures:

- If the gauge becomes damaged or if any other emergency or unusual situation arises:
    - Stop use of the gauge.

— Immediately secure the area and keep people away from the gauge until the situation is assessed and radiation levels are known. However, perform first aid for any injured individuals and remove them from the area only when medically safe to do so.

— If any equipment is involved, isolate the equipment until it is determined there is no contamination present.

— Gauge users and other potentially contaminated individuals should not leave the scene until emergency assistance arrives.

— Notify the persons in the order listed below of the situation:

NAME[2]	WORK PHONE NUMBER[2]	HOME PHONE NUMBER[2]
_____	_____	_____
_____	_____	_____
_____	_____	_____

- Follow the directions provided by the person contacted above.

RSO and Licensee Management:

- Arrange for a radiation survey to be conducted as soon as possible by a knowledgeable person using appropriate radiation detection instrumentation. This person could be a licensee employee using a survey meter, a local emergency responder or a consultant. To accurately assess the radiation danger, it is essential that the person performing the survey be competent in the use of the survey meter.

- Make necessary notifications to local authorities as well as the NRC as required. Appendix P contains typical NRC incident notifications required for fixed gauge licensees. (Even if not required to do so, you may report ANY incident to NRC by calling NRC's Operations Center at 301-816-5100 or 301-951-0550, which is staffed 24 hours a day and accepts collect calls.) NRC notification is required when gauges containing licensed material are lost or stolen and when gauges are damaged or involved in incidents that result in doses in excess of 10 CFR 20.2203 limits. Reporting requirements are found in 10 CFR 20.2201-2203 and 10 CFR 30.50.

---

[2]   Fill in with (and update, as needed) the names and telephone numbers of appropriate personnel (e.g., the RSO, AUs, or other knowledgeable licensee staff, licensee's consultant, gauge manufacturer, distributor or representative, fire department, or other emergency response organization, as appropriate, and the NRC) to be contacted in case of emergency.

Copies of operating and emergency procedures must be posted at each location of use or if posting procedures is not practicable, a notice which briefly describes the procedures and states where they may be examined may be posted instead.

Copies of operating and emergency procedures should be provided to all gauge users.

Appendix M

Model Leak Test Program

Model Leak Test Program

Training

Before allowing an individual to perform leak testing, the RSO will ensure that he or she has sufficient classroom and on-the-job training to show competency in performing leak tests independently.

Classroom training may be in the form of lecture, videotape, or self-study and will cover the following subject areas:

- Principles and practices of radiation protection

- Radioactivity measurements, monitoring techniques, and the use of instruments

- Mathematics and calculations basic to the use and measurement of radioactivity

- Biological effects of radiation.

Appropriate on-the-job-training consists of:

- Observing authorized personnel collecting and analyzing leak test samples

- Collecting and analyzing leak test samples under the supervision and in the physical presence of an individual authorized to perform leak tests.

Facilities and Equipment

- To ensure achieving the required sensitivity of measurements, leak tests will be analyzed in a low-background area.

- Individuals conducting leak tests will use a calibrated and operable survey instrument to check leak test samples for gross contamination before they are analyzed.

- A NaI(Tl) well counter system with a single or multichannel analyzer will be used to count samples from gauges containing gamma-emitters (e.g., Cs-137, Co-60).

- A liquid scintillation or gas-flow proportional counting system will be used to count samples from gauges containing beta-emitters (e.g., Sr-90) or alpha emitters (e.g., Am-241).

Frequency for Conducting Leak Tests of Sealed Sources

- Leak tests will be conducted at the frequency specified in the respective SSD Registration Certificate.

Procedure for Performing Leak Testing and Analysis

- For each source to be tested, list identifying information such as gauge serial number, radionuclide, activity.

- If available, use a survey meter to monitor exposure.

- Prepare a separate wipe sample (e.g., cotton swab or filter paper) for each source.

- Number each wipe to correlate with identifying information for each source.

- Wipe the most accessible area where contamination would accumulate if the sealed source were leaking.

- Select an instrument that is sensitive enough to detect 185 Bq (0.005 microcurie) of the radionuclide contained in the gauge.

- Using the selected instrument count and record background count rate.

- Check the instrument's counting efficiency using standard source of the same radionuclide as the source being tested or one with similar energy characteristics. Accuracy of standards should be within ± 5% of the stated value and traceable to a primary radiation standard such as those maintained by the National Institutes of Standards and Technology (NIST).

- Calculate efficiency.

For example:
$$\frac{[(\text{cpm from std}) - (\text{cpm from bkg})]}{\text{activity of std in Bq}} = \text{efficiency in cpm/Bq}$$

where: cpm = counts per minute
        std = standard
        bkg = background
        Bq = Becquerel

- Count each wipe sample; determine net count rate.

- For each sample, calculate and record estimated activity in Bq (or microcuries).

For example:
$$\frac{[(\text{cpm from wipe sample}) - (\text{cpm from bkg})]}{\text{efficiency in cpm/Bq}} = \text{Bq on wipe sample}$$

- Sign and date the list of sources, data and calculations. Retain records for 3 years.

- If the wipe test activity is 185 Bq (0.005 microcurie) or greater, notify the RSO, so that the source can be withdrawn from use and disposed of properly. Also notify NRC.

Reference: See the Notice of Availability (on the inside front cover of this report) to obtain a copy of Draft RG FC 412-4, "Guide for the Preparation of Applications for Licenses for the Use of Radioactive Materials in Leak-Testing Services," dated June 1985.

Appendix N

Information Needed to Support Applicant's Request to Perform Non-Routine Operations

Information Needed to Support Applicant's Request to Perform Non-Routine Operations

Applicants should review the section in this document on "Maintenance," which discusses, in general, licensee responsibilities before any maintenance or repair is performed.

Non-routine operations include installation of the gauge, initial radiation survey, repair or maintenance involving or potentially affecting components, including electronics, related to the radiological safety of the gauge (e.g., the source, source holder, source drive mechanism, shutter, shutter control, or shielding), gauge relocation, replacement, and disposal of sealed sources, alignment, removal of a gauge from service, and any other activities during which personnel could receive radiation doses exceeding NRC limits. See Figure 8.9.

Any non-manufacturer/non-distributor supplied replacement components or parts, or the use of materials (e.g., lubricants) other than those specified or recommended by the manufacturer or distributor need to be evaluated to ensure that they do not degrade the engineering safety analysis performed and accepted as part of the device registration. Licensees also need to ensure that, after maintenance or repair is completed, the gauge is tested and functions as designed, before the unit is returned to routine use.

If non-routine operations are not performed properly with attention to good radiation safety principles, the gauge may not operate as designed and personnel performing these tasks could receive radiation doses exceeding NRC limits. Radionuclides and activities in fixed gauges vary widely. For illustrative purposes in less than one minute, an unshielded cesium-137 source with an activity of 100 millicuries can deliver 0.05 Sv (5 rems) to a worker's hands or fingers (i.e., extremities), assuming the extremities are 1 centimeter from the source. However, gauges can contain sources of even higher activities with correspondingly higher dose rates. The threshold for extremity monitoring is 0.05 Sv (5 rems) per year.

Thus, applicants wishing to perform non-routine operations must use personnel with special training and follow appropriate procedures consistent with the manufacturer's or distributor•s instructions and recommendations that address radiation safety concerns (e.g., use of radiation survey meter, shielded container for the source, and personnel dosimetry (if required)). Accordingly, provide the following information:

Describe the types of work, maintenance, cleaning, repair that involve:

- Installation, relocation, or alignment of the gauge

- Components, including electronics, related to the radiological safety of the gauge (e.g., the source, source holder, source drive mechanism, shutter, shutter control, or shielding)

- Replacement and disposal of sealed sources

- Removal of a gauge from service

- A potential for any portion of the body to come into contact with the primary radiation beam; or

- Any other activity during which personnel could receive radiation doses exceeding NRC limits.

The principal reason for obtaining this information is to assist in the evaluation of the qualifications of individuals who will conduct the work and the radiation safety procedures they will follow.

A licensee may initially mount a gauge, without specific NRC or Agreement State authorization, if the gauge's SSD Certificate explicitly permits mounting of gauges by users and under the following conditions:

- The gauge must be mounted according to written instructions provided by the manufacturer or distributor;
- The gauge must be mounted in a location compatible with the "Conditions of Normal Use" and "Limitations and/or Other Considerations of Use" in the certificate of registration issued by NRC or an Agreement State;
- The on-off mechanism (shutter) must be locked in the off position, if applicable, or the source must be otherwise fully shielded;
- The gauge must be received in good condition (package was not damaged); and
- The gauge must not require any modification to fit in the proposed location.

Mounting does not include electrical connection, activation, or operation of the gauge. The source must remain fully shielded and the gauge may not be used until it is installed and made operational by a person specifically licensed by the Commission or an Agreement State to perform such operations.

- Identify who will perform non-routine operations and their training and experience. Acceptable training would include manufacturer's or distributor•s courses for non-routine operations or equivalent.

- Submit procedures for non-routine operations. These procedures should ensure the following:

  — doses to personnel and members of the public are within regulatory limits and ALARA (e.g., use of shielded containers or shielding);

  — the source is secured against unauthorized removal or access or under constant surveillance;

  — appropriate labels and signs are used;

  — manufacturer's or distributor•s instructions and recommendations are followed;

  — any non-manufacturer/non-distributor supplied replacement components or parts, or the use of materials (e.g., lubricants) other than those specified or recommended by the manufacturer or distributor are evaluated to ensure that they do not degrade the engineering safety analysis performed and accepted as part of the device registration; and

— before being returned to routine use, the gauge is tested to verify that it functions as designed and source integrity is not compromised.

- Confirm that individuals performing non-routine operations on gauges will wear both whole body and extremity monitoring devices or perform a prospective evaluation demonstrating that unmonitored individuals performing non-routine operations are not likely to receive, in one year, a radiation dose in excess of 10% of the allowable limits.

- Verify possession of at least one survey instrument that meets the criteria in "Radiation Safety Program - Instruments in NUREG-1556, Vol. 4, 'Consolidated Guidance about Materials Licenses: Program-Specific Guidance about Fixed Gauges Licenses,' dated October 1998."

- Describe steps to be taken to ensure that radiation levels in areas where non-routine operations will take place do not exceed 10 CFR 20.1301 limits. For example, applicants can do the following:

    — commit to performing surveys with a survey instrument (as described above);

    — specify where and when surveys will be conducted during non-routine operations; and

    — commit to maintaining, for 3 years from the date of the survey, records of the survey (e.g., who performed the survey, date of the survey, instrument used, measured radiation levels correlated to location of those measurements), as required by 10 CFR 20.2103.

Appendix O

Major DOT Regulations; Sample Shipping Documents, Placards and Labels

Major DOT Regulations; Sample Shipping Documents, Placards and Labels

The major areas in the DOT regulations that are most relevant for transportation of typical fixed gauges that are shipped as Type A quantities are as follows:

- Hazardous Materials Table, 49 CFR 172.101, Appendix A, list of hazardous substances and reportable quantities (RQ), Table 2: radionuclides

- Shipping Papers 49 CFR 172.200, 172.201, 172.202, 172.203, 172.204: general entries, description, additional description requirements, shipper's certification

- Package Markings 49 CFR 172.300, 172.301, 172.303, 172.304, 172.310, 172.324: General marking requirements for non-bulk packagings, prohibited marking, marking requirements, radioactive material, hazardous substances in non-bulk packaging

- Package Labeling 49 CFR 172.400, 172.401, 172.403, 172.406, 172.407, 172.436, 172.438, 172.440: General labeling requirements, prohibited labeling, radioactive materials, placement of labels, specifications for radioactive labels

- Placarding of Vehicles 49 CFR 172.500, 172.502, 172.504, 172.506, 172.516, 172.519, 172.556: Applicability, prohibited and permissive placarding, general placarding requirements, providing and affixing placards: highway, visibility and display of placards, specifications for RADIOACTIVE placards

- Emergency Response Information, Subpart G, 49 CFR 172.600, 172.602, 172.604: Applicability and general requirements, emergency response information, emergency response telephone number

- Training, Subpart H, 49 CFR 172.702, 172.704: Applicability and responsibility for training and testing, training requirements

- Radiation Protection Program for Shippers and Carriers, Subpart I, 49 CFR 172.801, 172.803, 172.805: Applicability of the radiation protection program, radiation protection program, recordkeeping, and notifications

- Shippers - General Requirements for Shipments and Packaging, Subpart I, 49 CFR 173.403, 173.410, 173.412, 173.415, 173.431, 173.433, 173.435, 173.441, 173.443, 173.448, 173.475, 173.476: Definitions, general design requirements, additional design requirements for Type A packages, authorized Type A packages, activity limits for Type A... packages, requirements for determining A1 and A2..., table of A1 and A2 values for radionuclides, radiation level limitations, contamination control, general transportation requirements, quality control requirements prior to each shipment, approval of special form radioactive materials

- Carriage by Public Highway - General Information and Regulations, Subpart A, 49 CFR 177.816, 177.817, 177.834(a), 177.842: Driver training, shipping paper, general requirements (secured against movement), Class 7 (radioactive) material.

Note: Type B shipping packages transport quantities of radionuclides greater than Type A allowable quantities. Requirements for Type B packages are in 10 CFR Part 71.

Hazard Communications for Class 7 (Radioactive) Materials

Marking Packages (49 CFR 172.300-338)

NOTE: IAEA, ICAO, and IMO may require additional hazard communication information for international shipments
This table must not be used as a substitute for the DOT and NRC regulations on the transportation of radioactive materials

Markings Always Required Unless Excepted	Additional Markings Sometimes Required	Optional Markings
<u>Non-Bulk Packages</u> • Proper shipping name • U.N. identification number • Name and address of consignor or consignee, *unless*: - highway only and no motor carrier transfers, <u>or</u> - part of carload or truckload lot or freight container load, and entire contents of railcar, truck, or freight container are shipped from one consignor to one consignee [see §172.301(d)] - - - - - - - - - - <u>Bulk Packages</u> (i.e., net capacity greater than 119 gallons as a receptacle for liquid, or 119 gallons and 882 pounds as a receptacle for solid, or water capacity greater than 1000 lbs, with no consideration of intermediate forms of containment) • U.N. identification number, on orange, rectangular panel (see §172.332) - some exceptions exist	<u>Materials-Based Requirements:</u> • If in excess of 110 lbs (50 kg), Gross Weight • If non-bulk <u>liquid</u> package, underlined double arrows indicating upright orientation (two opposite sides) [ISO Std 780-1985 marking] • If a Hazardous substance in non-bulk package, the letters "RQ" in association with the proper shipping name <u>Package-Based Requirements:</u> • The package type if Type A or Type B (½" or greater letters) • The specification-required markings [e.g., for Spec. 7A packages: "DOT 7A Type A" and "Radioactive Material" (see §178.350-353)] • For approved packages, the certificate ID number (e.g., USA/9166/B(U), USA/9150/B(U)-85, ...) • If Type B, the trefoil (radiation) symbol per Part 172 App. B [*size*: outer radius ≥ 20 mm (0.8 in)] • For NRC certified packages, the model number, gross weight, and package ID number (10 CFR 71.85) <u>Administrative-Based Requirements:</u> • If a DOT exemption is being used, "DOT-E" followed by the exemption number • If an export shipment, "USA" in conjunction with the specification markings or certificate markings	• "IP-1," "IP-2," or "IP-3" on industrial packaging is recommended • Both the name and address of consignor and consignee are recommended • Other markings (e.g., advertising) are permitted, but must be sufficiently away from required markings and labeling

Some Special Considerations/Exceptions for Marking Requirements

• Marking is required to be: (1) durable, (2) printed on a package, label, tag, or sign, (3) unobscured by labels or attachments, (4) isolated from other marks, and (5) be representative of the hazmat contents of the package

• Limited Quantity (§173.421) packages and Articles Containing Natural Uranium and Thorium (§173.426) must bear the marking "radioactive" on the outside of the inner package or the outer package itself, and are excepted from other marking. The excepted packages shipped under UN 2910 must also have the accompanying statement that is required by §173.422.

• Empty (§173.428) and Radioactive Instrument and Article (§173.424) packages are excepted from marking

• Shipment of LSA or SCO required by §173.427 to be consigned as exclusive use are excepted from marking except that the exterior of each nonbulk package must be marked **"Radioactive-LSA"** or **"Radioactive-SCO,"** as appropriate. Examples of this category are domestic, strong-tight containers with less than an A_2 quantity, and domestic NRC certified LSA/SCO packages using 10 CFR 71.52.

• For bulk packages, marking may be required on more than one side of the package (see 49 CFR 172.302(a))

Hazard Communications for Class 7 (Radioactive) Materials

DOT Shipping Papers (49 CFR 172.200-205)

NOTE: IAEA, ICAO, and IMO may require additional hazard communication information for international shipments
This table must not be used as a substitute for the DOT and NRC regulations on the transportation of radioactive materials

Entries Always Required Unless Excepted	Additional Entries Sometimes Required	Optional Entries
• The basic description, In sequence: **Proper Shipping Name, Hazard Class (7), U.N. Identification Number** • 24 hour **emergency response telephone number** • Name of **shipper** • Proper page numbering (Page 1 of 4) • Except for empty and bulk packages, the **total quantity** (mass, or volume for liquid), in appropriate units (lbs, mL....) • If not special form, **chemical and physical form** • The **name of each radionuclide** (95% rule) and total package activity. The activity must be in SI units (e.g., Bq, TBq), or both SI units and customary units (e.g., Ci, mCi). However, for <u>domestic shipments</u>, the activity *may* be expressed in terms of customary units only, until 4/1/97. • For each labeled package: - The **category of label** used; - The **transport index** of each package with a Yellow-II or Yellow-III label • Shipper's **certification** (not required of private carriers)	<u>Materials-Based Requirements:</u> • If hazardous substance, "RQ" as part of the basic description • The LSA or SCO group (e.g., LSA-II) • "Highway Route Controlled Quantity" as part of the basic description , if HRCQ • Fissile material information (e.g., "Fissile Exempt," controlled shipment statement [see §172.203(d)(7)]) • If the material is considered hazardous waste and the word waste does not appear in the shipping name, then "waste" must precede the shipping name (e.g., Waste Radioactive Material, nos, UN2982) • "Radioactive Material" if not in proper shipping name <u>Package-Based Requirements:</u> • Package identification for DOT Type B or NRC certified packages • IAEA CoC ID number for export shipments or shipments using foreign-made packaging (see §173.473) <u>Administrative-Based Requirements:</u> • "Exclusive Use-Shipment" • Instructions for maintenance of exclusive use-shipment controls for LSA/SCO strong-tight or NRC certified LSA (§ 173.427) • If a DOT exemption is being used, "DOT-E" followed by the exemption number	• The type of packaging (e.g., Type A, Type B, IP-1, ...) • The Technical/chemical name may be in included (if listed in §172.203(k), in parentheses between the proper shipping name and hazard class; otherwise inserted in parenthesis after the basic description) • Other information is permitted (e.g., functional description of the product), provided it does not confuse or detract from the proper shipping name or other required information • For fissile radionuclides, except Pu-238, Pu-239, and Pu-241, the weight in grams or kilograms may be used *in place of* activity units. For Pu-238, Pu-239, and Pu-241, the weight in grams or kilograms may optionally be entered *in addition to* activity units [see § 172.203(d)(4)] • Emergency response hazards and guidance information (§§ 172.600-604) may be entered on the shipping papers, or may be carried with the shipping papers [§ 172.602(b)]

Some Special Considerations/Exceptions for Shipping Paper Requirements

• Shipments of Radioactive Material, excepted packages, under UN2910 (e.g., Limited Quantity, Empty packages, and Radioactive Instrument and Article), are excepted from shipping papers. For limited quantities (§173.421), this is only true if the limited quanttiy is not a hazardous substance (RQ) or hazardous waste (40 CFR 262)

• Shipping papers must be in the pocket on the left door, or readily visible to person entering driver's compartment and within arm's reach of the driver

• For shipments of multiple cargo types, any HAZMAT entries must appear as the first entries on the shipping papers, be designated by an "X" (or "RQ") in the hazardous material column, <u>or</u> be highlighted in a contrasting color

NRC Contacts: John Cook, (301) 415-8521   Earl Easton, (301) 415-8520

Hazard Communications for Class 7 (Radioactive) Materials

Labeling Packages (49 CFR 172.400-450)

NOTE: IAEA, ICAO, and IMO may require additional hazard communication information for international shipments
This table must not be used as a substitute for the DOT and NRC regulations on the transportation of radioactive materials

Placement of Radioactive Labels

- Labeling is required to be: (1) placed near the required marking of the proper shipping name, (2) printed or affixed to the package surface (not the bottom), (3) in contrast with its background, (4) unobscured by markings or attachments, (5) within color, design, and size tolerance, and (6) representative of the HAZMAT contents of the package

- For labeling of radioactive materials packages, two labels are required on opposite sides excluding the bottom

Determination of Required Label

Size:				
Sides: ≥ 100 mm (3.9 in.) *Border:* 5-6.3 mm (0.2-0.25 in.)	49 CFR 172.436	49 CFR 172.438	49 CFR 172.440	EMPTY 49 CFR 172.450
Label	**WHITE-I**	**YELLOW-II**	**YELLOW-III**	**EMPTY LABEL**
Required when:	Surface radiation level < 0.005 mSv/hr (0.5 mrem/hr)	0.005 mSv/hr (0.5 mrem/hr) < surface radiation level ≤ 0.5 mSv/hr (50 mrem/hr)	0.5 mSv/hr (50 mrem/hr) < surface radiation level < 2 mSv/hr (200 mrem/h) [Note: 10 mSv/hr (1000 mrem/hr) for exclusive-use closed vehicle (§173.441(b)]	The EMPTY label is required for shipments of empty Class 7 (radioactive) packages made pursuant to §173.428. It must cover any previous labels, or they must be removed or obliterated.
Or:	TI = 0 [1 meter dose rate < 0.0005 mSv/hr (0.05 mrem/hr)]	TI ≤ 1 [1 meter dose rate < 0.01 mSv/hr (1 mrem/hr)]	TI < 10 [1 meter dose rate < 0.1 mSv/hr (10 mrem/hr)] [Note: There is no *package* TI limit for exclusive-use]	

Notes:
- Any package containing a Highway Route Controlled Quantity (HRCQ) must bear YELLOW-III label
- Although radiation level transport indices (TIs) are shown above, for fissile material, the TI is typically determined on the basis of criticality control

Content on Radioactive Labels

- RADIOACTIVE Label must contain (entered using a durable, weather-resistant means):
  (1) The radionuclides in the package (with consideration of available space). Symbols (e.g., Co-60) are acceptable
  (2) The activity in SI units (e.g., Bq, TBq), or both SI units with customary units (e.g., Ci, mCi) in parenthesis. However, for domestic shipments, the activity *may* be expressed in terms of customary units only, until 4/1/97.
  (3) The Transport Index (TI) in the supplied box. The TI is entered *only* on YELLOW-II and YELLOW-III labels

Some Special Considerations/Exceptions for Labeling Requirements

- For materials meeting the definition of another hazard class, labels for each secondary hazard class need to be affixed to the package. The subsidiary label *may* not be required on opposite sides, and must not display the hazard class number

- Radioactive Material, excepted packages, under UN2910 (e.g., Limited Quantity, Empty packages, and Radioactive Instrument and Article), are excepted from labeling. However, if the excepted quantity meets the definition for another hazard class, it is re-classed for that hazard. Hazard communication requirements for the other class are required

- Labeling exceptions exist for shipment of LSA or SCO required by § 173.427 to be consigned as exclusive use

- The "Cargo Aircraft Only" label is typically required for radioactive materials packages shipped by air [§ 172.402(c)]

Hazard Communications for Class 7 (Radioactive) Materials

Placarding Vehicles (49 CFR 172.500-560)
NOTE: IAEA, ICAO, and IMO may require additional hazard communication information for international shipments
This table must not be used as a substitute for the DOT and NRC regulations on the transportation of radioactive materials

Visibility and Display of Radioactive Placard

- Placards are required to be displayed:
  - on four sides of the vehicle
  - visible from the direction they face, (for the front side of trucks, tractor-front, trailer, or both are authorized)
  - clear of appurtenances and devices (e.g., ladders, pipes, tarpaulins)
  - at least 3 inches from any markings (such as advertisements) which may reduce placard's effectiveness
  - upright and on-point such that the words read horizontally
  - in contrast with the background, or have a lined-border which contrasts with the background
  - such that dirt or water from the transport vehicle's wheels will not strike them
  - securely attached or affixed to the vehicle, or in a holder.
- Placard must be maintained by carrier to keep color, legibility, and visibility.

Conditions Requiring Placarding

- Placards are required for any vehicle containing package with a RADIOACTIVE Yellow-III label
- Placards are required for shipment of LSA or SCO required by §173.427 to be consigned as exclusive use. Examples of this category are domestic, strong-tight containers with less than an A_2 quantity, and domestic NRC certified LSA/SCO packages using 10 CFR 71.52. Also, for bulk packages of these materials, the orange panel marking with the UN Identification number is not required.
- Placards are required any vehicle containing package with a Highway Route Controlled Quantity (HRCQ). In this case, the placard must be placed in a square background as shown below (see §173.507(a))

Radioactive Placard

Size Specs:

Sides: $\geq$ 273 mm (10.8 in.)

Solid line Inner border: About 12.7 mm (0.5 in.) from edges

Lettering: $\geq$ 41 mm (1.6 in.)

Square for HRCQ: 387mm (15.25 in.) outside length by 25.4 mm (1 in.) thick

49 CFR 172.556

RADIOACTIVE PLACARD (Domestic)

Base of yellow solid area: 29 ± 5 mm (1.1 + 0.2 in.) above horizontal centerline

IAEA SS 6 (1985) paras. 443-444

RADIOACTIVE PLACARD (International)

See 49 CFR 172.527 AND 556

RADIOACTIVE PLACARD FOR HIGHWAY ROUTE CONTROLLED QUANTITY
(either domestic or international placard could be in middle)

Some Special Considerations/Exceptions for Placarding Requirements

- Domestically, substitution of the UN ID number for the word "RADIOACTIVE" on the placard is prohibited for Class 7 materials. However, some import shipments may have this substitution in accordance with international regulations.
- Bulk packages require the orange, rectangular panel marking containing the UN ID number, which must be placed adjacent to the placard (see §172.332) [NOTE: except for LSA/ SCO exclusive use under §173.427, as above]
- If placarding for more than one hazard class, subsidiary placards must not display the hazard class number. Uranium Hexaflouride (UF_6) shipments $\geq$ 454 kg (1001 lbs) require both RADIOACTIVE and CORROSIVE (Class 8) placarding
- For shipments of radiography cameras in convenience overpacks, if the overpack does not require a RADIOACTIVE - YELLOW III label, vehicle placarding is not required (regardless of the label which must be placed on the camera)

Minimum Required Packaging For Class 7 (Radioactive) Materials

This table must not be used as a substitute for the DOT and NRC regulations on the transportation of radioactive materials

Quantity:	< 70 Bq/g (< 0.002 · Ci/g)	Limited Quantity (§173.421)	A_1/A_2 value	1 rem/hr at 3 m, unshielded (§173.435) (§173.427)
Non-LSA/SCO:		Excepted	Type A	Type B [3]
Domestic or International LSA/SCO: LSA-I solid, (liquid)[1] SCO-I	Excepted		IP-I	Type B [3]
LSA-I Liquid LSA-II Solid, (liquid or gas)[1] (LSA-III)[1] SCO-II			IP-II	Type B [3]
LSA-II Liquid or Gas LSA-III			IP-III	Type B [3]
Domestic (only) LSA/SCO: LSA-I, II, III; SCO-I, II		Excepted	Strong-tight [2]	DOT Spec. 7A Type A → Type B [3] / NRC Type A LSA [3,4]

1. For entries in parentheses, exclusive use is required for shipment in an IP (e.g., shipment of LSA-I liquid in an IP-I packaging would require exclusive use consignment)
2. Exclusive use required for strong-tight container shipments made pursuant to §173.427(b)(2)
3. Subject to conditions in Certificate, if NRC package
4. Exclusive use required, see §173.427(b)(4). Use of these packages expires on 4/1/99 (10 CFR 71.52)

Package and Vehicle Radiation Level Limits (49 CFR 173.441) [A]

This table must not be used as a substitute for the DOT and NRC regulations on the transportation of radioactive materials

Transport Vehicle Use:	Non-Exclusive	Exclusive		
Transport Vehicle Type:	Open or Closed	Open (flat-bed)	Open w/Enclosure [B]	Closed
Package (or freight container) Limits:				
External Surface	2 mSv/hr (200 mrem/hr)	2 mSv/hr (200 mrem/hr)	10 mSv/hr (1000 mrem/hr)	10 mSv/hr (1000 mrem/hr)
Transport Index (TI) [C]	10	no limit		
Roadway or Railway Vehicle (or freight container) Limits:				
Any point on the outer surface	N/A	N/A	N/A	2 mSv/hr (200 mrem/hr)
Vertical planes projected from outer edges		2 mSv/hr (200 mrem/hr)	2 mSv/hr (200 mrem/hr)	N/A
Top of . . .		load: (200 mrem/hr)	enclosure: 2 mSv/hr (200 mrem/hr)	vehicle: 2 mSv/hr (200 mrem/hr)
2 meters from. . .		vertical planes: 0.1 mSv/hr (10 mrem/hr)	vertical planes: 0.1 mSv/hr (10 mrem/hr)	outer lateral surfaces: 0.1 mSv/hr (10 mrem/hr)
Underside		2 mSv/hr (200 mrem/hr)		
Occupied position	N/A [D]	0.02 mSv/hr (2 mrem/hr) [E]		
Sum of package TI's	50	no limit [F]		

A. The limits in this table do not apply to excepted packages - see 49 CFR 173.421-426
B. Securely attached (to vehicle), access-limiting enclosure; package personnel barriers are considered as enclosures
C. For nonfissile radioactive materials packages, the dimensionless number equivalent to maximum radiation level at 1 m (3.3 feet) from the exterior package surface, in millirem/hour
D. No dose limit is specified, but separation distances apply to Radioactive Yellow-II or Radioactive Yellow-III labeled packages
E. Does not apply to private carrier wearing dosimetry if under radiation protection program satisfying 10 CFR 20 or 49 CFR 172 Subpart I
F. Some fissile shipments may have combined conveyance TI limit of 100 - see 10 CFR 71.59 and 49 CFR 173.457

Package and Vehicle Contamination Limits (49 CFR 173.443)	
This table must not be used as a substitute for the DOT and NRC regulations on the transportation of radioactive materials	

NOTE: All values for contamination in DOT rules are to be averaged over each 300 cm^2
Sufficient measurements must be taken in the appropriate locations to yield representative assessments

• • means the sum of beta emitters, gamma emitters, and low-toxicity alpha emitters
"• means the sum of all other alpha emitters (i.e., other than low-toxicity alpha emitters)

The Basic Contamination Limits for All Packages: 49 CFR 173.443(a), Table 11	General Requirement: Non-fixed (removable) contamination must be kept as low as reasonably achievable (ALARA)
	• • : 0.4 Bq/cm^2 = 40 Bq/100 cm^2 = 1x10^{-5} • Ci/cm^2 = 2200 dpm/100 cm^2
	" : 0.04 Bq/cm^2 = 4 Bq/100 cm^2 = 1x10^{-6} • Ci/cm^2 = 220 dpm/100 cm^2

The following exceptions and deviations from the above basic limits exist:

Deviation from Basic Limits	Regulation 49 CFR §§	Applicable Location and Conditions Which must Be Met:
10 times the basic limits	173.443(b) and 173.443(c) Also see 177.843 (highway)	On any external surface of a package in an exclusive use shipment, during transport including end of transport. Conditions include: (1) Contamination levels at beginning of transport must be below the basic limits. (2) Vehicle must not be returned to service until radiation level is shown to be ≤ 0.005 mSv/hr (0.5 mrem/hr) at any accessible surface, and there is no significant removable (non-fixed) contamination.
10 times the basic limits	173.443(d) Also see 177.843 (highway)	On any external surface of a package, at the beginning or end of transport, if a closed transport vehicle is used, solely for transporting radioactive materials packages. Conditions include: (1) A survey of the interior surfaces of the empty vehicle must show that the radiation level at any point does not exceed 0.1 mSv/hr (10 mrem/hr) at the surface, or 0.02 mSv/hr (2 mrem/hr) at 1 meter (3.3 ft). (2) Exterior of vehicle must be conspicuously stenciled, "For Radioactive Materials Use Only" in letters at least 76 mm (3 inches) high, on both sides. (3) Vehicle must be kept closed except when loading and unloading.
100 times the basic limits	173.428	Internal contamination limit for excepted package-empty packaging, Class 7 (Radioactive) Material, shipped in accordance with 49 CFR 173.428. Conditions include: (1) The basic contamination limits (above) apply to **external** surfaces of package. (2) Radiation level must be ≤ 0.005 mSv/hr (0.5 mrem/hr) at any external surface. (3) Notice in §173.422(a)(4) must accompany shipment. (4) Package is in unimpaired condition & securely closed to prevent leakage. (5) Labels are removed, obliterated, or covered, and the "empty" label (§172.450) is affixed to the package.

In addition, after any incident involving spillage, breakage, or suspected contamination, the modal-specific DOT regulations (§177.861(a), highway; §174.750(a), railway; and §175.700(b), air) specify that vehicles, buildings, areas, or equipment have "no significant removable surface contamination," before being returned to service or routinely occupied. The carrier must also notify offeror at the earliest practicable moment after incident.

STRAIGHT BILL OF LADING
ORIGINAL—NOT NEGOTIABLE

Appendix P

Shipper No. _____

Carrier No. _____

Page __1__ of __1__

(Name of carrier)   (SCAC)   Date _____

TO: Consignee	ABC Bottling, Inc. Chicago Plant	FROM: Shipper	ABC Bottling, Inc. Milwaukee Plant

On Collect on Delivery shipments, the letters "COD" must appear before consignee's name or as otherwise provided in Item 430, Sec. 1.

Street	Admiral Avenue	Street	Liberty Place
Destination	Chicago, Illinois Zip Code 35011	Origin	Milwaukee, Wisconsin 38023

Route

Vehicle Number

No. of Units & Container Type	HM	BASIC DESCRIPTION Proper Shipping Name, Hazard Class Identification Number (UN or NA) per 172.101, 172 202, 172 203	TOTAL QUANTITY (Weight, Volume, Gallons, etc.)	WEIGHT (Subject to Correction)	RATE	CHARGES (For Carrier Use Only)
1	RQ	Radioactive material, special form,				
		n.o.s. 7 UN 2974				
		55.5 GBq (1.5 Ci) Cs-137	55.5 GBq			
			(1.5 Ci)			
		RADIOACTIVE - YELLOW II				
		TI = 0.4 **				
		USDOT 7A TYPE A				
		Emergency Response Telephone No.: 1-800-000-0000	(24 hr/d)**			
		** SUBSTITUTE APPROPRIATE INFORMATION	FOR YOUR			
		GAUGE AND SHIPMENT				

PLACARDS TENDERED: YES ☐ NO ☐

REMIT C.O.D. TO: ADDRESS

Note — Where the rate is dependent on value, shippers are required to state specifically in writing the agreed or declared value of the property.

The agreed or declared value of the property is hereby specifically stated by the shipper to be not exceeding

$ _____ per _____

I hereby declare that the contents of this consignment are fully and accurately described above by proper shipping name and are classified, packaged, marked and labeled, and are in all respects in proper condition for transport by ☒ Highway ☐ (DELETE NON-APPLICABLE MODE OF TRANSPORT) according to applicable international and national governmental regulations.

John Jones   Signature

COD   Amt: $ _____

Subject to Section 7 of the conditions, if this shipment is to be delivered to the consignee without recourse on the consignor, the consignor shall sign the following statement:
The carrier shall not make delivery of this shipment without payment of freight and all other lawful charges.

_____ (Signature of Consignor)

C.O.D. FEE: PREPAID ☐ COLLECT ☐	$
TOTAL CHARGES	$

FREIGHT CHARGES
FREIGHT PREPAID
except when box at right is checked ☐
Check box if charges are to be collect

RECEIVED, subject to the classifications and lawfully filed tariffs in effect on the date of the issue of this Bill of Lading, the property described above in apparent good order, except as noted (contents and condition of contents of packages unknown), marked, consigned, and destined as indicated above which said carrier (the word carrier being understood throughout the contract as meaning any person or corporation in possession of the property under the contract) agrees to carry to its usual place of delivery at said destination, if on its route, otherwise to deliver to another carrier on the route to said destination. It is mutually agreed as to each carrier of all or any of said property over all or any portion of

said route to destination and as to each party at any time interested in all or any said property, that every service to be performed hereunder shall be subject to all the bill of lading terms and conditions in the governing classification on the date of shipment.
Shipper hereby certifies that he is familiar with all the bill of lading terms and conditions in the governing classification and the said terms and conditions are hereby agreed to by the shipper and accepted for himself and his assigns.

SHIPPER		CARRIER	
PER		PER	
		DATE	

1

Permanent post-office address of shipper:.

STYLE F65 LABELMASTER, Div. of American Labelmark Co., Chicago, IL 60646 312/478-0900

Appendix P

NRC Incident Notifications

NRC Incident Notifications

Table P.1 Typical NRC Incident Notifications Required for Fixed Gauge Licensees

Event	Telephone Notification	Written Report	Regulatory Requirement
Theft or loss of material	immediate	30 days	10 CFR 20.2201(a)(1)(i)
Whole body dose greater than 0.25 Sv (25 rems)	immediate	30 days	10 CFR 20.2202(a)(1)(i)
Extremity dose greater than 2.5 Sv (250 rems)	immediate	30 days	10 CFR 20.2202(a)(1)(iii)
Whole body dose greater than 0.05 Sv (5 rems) in 24 hours	24 hours	30 days	10 CFR 20.2202(b)(1)(i)
Extremity dose greater than 0.5 Sv (50 rems) in 24 hours	24 hours	30 days	10 CFR 20.2202(b)(1)(iii)
Whole body dose greater than 0.05 Sv (5 rems)	none	30 days	10 CFR 20.2203(a)(2)(i)
Dose to individual member of public greater than 1 mSv (100 mrems)	none	30 days	10 CFR 20.2203(a)(2)(iv)
Defect in equipment that could create a substantial safety hazard	2 days	30 days	10 CFR 21.21(d)(3)(i)
Filing petition for bankruptcy under 11 U.S.C.	none	immediately after filing petition	10 CFR 30.34(h)
Expiration of license	none	60 days	10 CFR 30.36(d)
Decision to permanently cease licensed activities at *entire site*	none	60 days	10 CFR 30.36(d)
Decision to permanently cease licensed activities in any *separate building or outdoor area* that is unsuitable for release for unrestricted use	none	60 days	10 CFR 30.36(d)
No principal activities conducted for 24 months *at the entire site*	none	60 days	10 CFR 30.36(d)
No principal activities conducted for 24 months *in any separate building or outdoor area* that is unsuitable for release for unrestricted use	none	60 days	10 CFR 30.36(d)

Event	Telephone Notification	Written Report	Regulatory Requirement
Event that prevents immediate protective actions necessary to avoid exposure to radioactive materials that could exceed regulatory limits	immediate	30 days	10 CFR 30.50(a)
Equipment is disabled or fails to function as designed when required to prevent radiation exposure in excess of regulatory limits	24 hours	30 days	10 CFR 30.50(b)(2)
Unplanned fire or explosion that affects the integrity of any licensed material or device, container, or equipment with licensed material	24 hours	30 days	10 CFR 30.50(b)(4)

Note: Telephone notifications shall be made to the NRC Operations Center at 301-816-5100 or 301-951-0550.

Appendix Q

Sample Fixed Gauge License

Sample Fixed Gauge License

A sample Fixed Gauge License appears on the following pages.

APPENDIX Q

U.S. NUCLEAR REGULATORY COMMISSION

MATERIALS LICENSE

Pursuant to the Atomic Energy Act of 1954, as amended, the Energy Reorganization Act of 1974 (Public Law 93-438), and Title 10, Code of Federal Regulations, Chapter I, Parts 30, 31, 32, 33, 34, 35, 36, 39, 40, and 70, and in reliance on statements and representations heretofore made by the licensee, a license is hereby issued authorizing the licensee to receive, acquire, possess, and transfer byproduct, source, and special nuclear material designated below; to use such material for the purpose(s) and at the place(s) designated below; to deliver or transfer such material to persons authorized to receive it in accordance with the regulations of the applicable Part(s). This license shall be deemed to contain the conditions specified in Section 183 of the Atomic Energy Act of 1954, as amended, and is subject to all applicable rules, regulations, and orders of the Nuclear Regulatory Commission now or hereafter in effect and to any conditions specified below.

Licensee	In accordance with the letter dated [insert the month, day and year],
1. Fixed Gauge Measurements, Inc.	3. License number 08-00000-01 is amended in its entirety to read as follows:
2. 1234 A Street, NW	4. Expiration date [Insert a date, last day of the month, 10 years after issuance date]
Washington, D.C. 20001	5. Docket No. 030-00000 Reference No.

6. Byproduct, source, and/or special nuclear material	7. Chemical and/or physical form	8. Maximum amount that licensee may possess at any one time under this license
A. Cobalt-60	A. Sealed source [insert manufacturer name] Model [insert model number]	A. No single source to exceed the maximum activity specified in the certificate of registration issued by the U.S. Nuclear Regulatory Commission or an Agreement State
B. Cesium-137	B. Sealed source [insert manufacturer name] Model [insert model number]	B. No single source to exceed the maximum activity specified in the certificate of registration issued by the U.S. Nuclear Regulatory Commission or an Agreement State
C. Americium-241	C. Sealed neutron source [insert manufacturer name] Model [insert model number]	C. No single source to exceed the maximum activity specified in the certificate of registration issued by the U.S. Nuclear Regulatory Commission or an Agreement State

NOTE: SPECIFY IN ITEM 7 ABOVE, THE MANUFACTURER AND MODEL NUMBER FOR EACH TYPE OF GAUGE REQUESTED.

License Number
08-00000-01

Docket or Reference Number
030-00000

Amendment No. XX

MATERIALS LICENSE
SUPPLEMENTARY SHEET

9. Authorized use:

   A. through C.   For [specify the use of the gauge], in fixed gauging devices in accordance with the certificate of registration issued by the U.S. Nuclear Regulatory Commission under 10 CFR 32.210 or with an Agreement State and which have been distributed in accordance with a Commission or Agreement State specific license authorizing distribution to persons specifically authorized by a Commission or Agreement State license to receive, possess, and use the devices.

CONDITIONS

10. Licensed material may be used only at the licensee's facilities located at [fill in the street address of the facility(s)].

11. Licensed material shall be used by, or under the supervision of individuals who have received the training described in [letter, application] dated [fill in date]. The licensee shall maintain records of individuals designated as suers for 3 years following the last use of licensed material by the individual.

12. The Radiation Safety Officer (RSO) for this license is [insert name of RSO]. **(This condition is used on licenses where NRC reviews the qualifications of the RSO on a case-by-case basis. The licensee would be charged an amendment fee to change the RSO.)**

OR

12. A.   The Radiation Safety Officer (RSO) for this license is [insert name of RSO].

   B.   Before assuming the duties and responsibilities as RSO for this license, the individual shall have successfully completed one of the training courses described in Criteria in Section 8.7.1 of NUREG-1556, Volume 4, dated October 1998. **(This condition is used on licenses where the licensee has committed to train the RSO as described in NUREG - 1556, Volume 4. No technical review would be needed for change of RSO, so the licensee would not be charged an amendment fee. The licensee must still notify NRC of the name of the new RSO.)**

13. A.   Sealed sources shall be tested for leakage and/or contamination at intervals not to exceed the intervals specified in the certificate of registration issued by the U.S. Nuclear Regulatory Commission under 10 CFR 32.210 or by an Agreement State.

   B.   Notwithstanding Paragraph A of this condition, sealed sources designed to primarily emit alpha particles shall be tested for leakage and/or contamination at intervals not to exceed 3 months.

NRC FORM 374A	U.S. NUCLEAR REGULATORY COMMISSION	PAGE 3 of 6 PAGES
	MATERIALS LICENSE **SUPPLEMENTARY SHEET**	License Number 08-00000-01
		Docket or Reference Number 030-00000
		Amendment No. XX

C.   In the absence of a certificate from a transferor indicating that a leak test has been made within the intervals specified in the certificate of registration issued by the U.S. Nuclear Regulatory Commission under 10 CFR 32.210 or by an Agreement State, prior to the transfer, a sealed source received from another person shall not be put into use until tested and the test results received.

D.   Sealed sources need not be tested if they contain only hydrogen-3; or they contain only a radioactive gas; or the half-life of the isotope is 30 days or less; or they contain not more than 100 microcuries of beta and/or gamma emitting material or not more than 10 microcuries of alpha emitting material.

E.   Sealed sources need not be tested if they are in storage and are not being used. However, when they are removed from storage for use or transferred to another person, and have not been tested within the required leak test interval, they shall be tested before use or transfer. No sealed source shall be stored for a period of more than 10 years without being tested for leakage and/or contamination.

F.   The leak test shall be capable of detecting the presence of 0.005 microcurie of radioactive material on the test sample. If the test reveals the presence of 0.005 microcurie or more of removable contamination, a report shall be filed with the U.S. Nuclear Regulatory Commission in accordance with 10 CFR 30.50(b)(2), and the source shall be removed immediately from service and decontaminated, repaired, or disposed of in accordance with Commission regulations. The report shall be filed within 5 days of the date the leak test result is known with the appropriate U.S. Nuclear Regulatory Commission, Regional Office referenced in Appendix D of 10 CFR Part 20. The report shall specify the source involved, the test results, and corrective action taken.

G.   Tests for leakage an/or contamination, limited to leak test sample collection shall be performed by persons specifically licensed by the U.S. Nuclear Regulatory Commission or an Agreement State to perform such services. The licensee is not authorized to perform the analysis. Analysis of leak test samples must be performed by persons specifically licensed by the Commission or an Agreement State to perform such services. **(This condition is used for licensees NOT authorized to perform leak test analysis.)**

OR

G.   Tests for leakage and/or contamination, including leak test sample collection and analysis, shall be performed by the licensee or other persons specifically licensed by the U.S. Nuclear Regulatory Commission or an Agreement State to perform such services. **(This condition is used for licensees authorized to collect AND analyze leak test samples.)**

H.   Records of leak test results shall be kept in units of microcuries and shall be maintained for 3 years.

NRC FORM 374A	U.S. NUCLEAR REGULATORY COMMISSION	PAGE 4 of 6 PAGES
	MATERIALS LICENSE **SUPPLEMENTARY SHEET**	License Number 08-00000-01
		Docket or Reference Number 030-00000
		Amendment No. XX

14. Sealed sources containing licensed material shall not be opened or sources removed from source holders by the licensee, except as specifically authorized.

15. The licensee shall conduct a physical inventory every 6 months, or at other intervals approved by the U.S. Nuclear Regulatory Commission, to account for all sealed sources and/or devices received and possessed under the license. Records of inventories shall be maintained for 5 years from the date of each inventory, and shall include the radionuclides, quantities, manufacturer's name and model numbers, and the date of the inventory.

16. A. Each gauge shall be tested for the proper operation of the on-off mechanism (shutter) and indicator, if any, at intervals not to exceed 6 months or at such longer intervals as specified in the certificate of registration issued by the U.S. Nuclear Regulatory Commission pursuant to 10 CFR 32.210 or the equivalent regulations of an Agreement State.

    B. Notwithstanding the periodic on-off mechanism (shutter) and indicator test, the requirement does not apply to gauges that are stored, not being used, and have the shutter lock mechanism in a locked position. The gauges exempted from this periodic test shall be tested before use.

17. The following services shall not be performed by the licensee: installation, initial radiation surveys, relocation, removal from service, dismantling, alignment, replacement, disposal of the sealed source and non-routine maintenance or repair of components related to the radiological safety of the gauge (i.e., the sealed source, the source holder, source drive mechanism, on-off mechanism (shutter), shutter control, shielding). These services shall be performed only by persons specifically licensed by the U.S. Nuclear Regulatory Commission or an Agreement State to perform such services. **(This condition is used when the licensee is NOT authorized to perform any non-routine operations.)**

OR

17. A. [Insert the services authorized in this license, delete remaining: installation, initial radiation surveys, relocation, removal from service dismantling, alignment, replacement, disposal of the sealed source and non-routine maintenance or repair of components related to the radiological safety of the gauge] shall be performed only by, [insert name(s)] or other individuals who have completed the training specified in [insert application/letter date] or by persons specifically licensed by the U.S. Nuclear Regulatory Commission or an Agreement State to perform such services. **(Part A of this condition is used to specify which non-routine operations may be performed by a licensee.)**

    B. The following services shall not be performed by the licensee: [insert the services NOT authorized in this license, delete remaining: installation, initial radiation surveys, relocation, removal from service, dismantling, alignment, replacement, disposal of the sealed sources and non-routine maintenance or repair of components related to the radiological safety of the gauge]. These services shall be performed only by persons specifically licensed by the U.S. Nuclear Regulatory

NRC FORM 374A	U.S. NUCLEAR REGULATORY COMMISSION	PAGE 5 of 6 PAGES

MATERIALS LICENSE
SUPPLEMENTARY SHEET

License Number
08-00000-01

Docket or Reference Number
030-00000

Amendment No. XX

Commission or an Agreement State to perform such services. **(Part B of the condition is used to specify which non-routine operations ia licensee is NOT authorized to perform.)**

18. The licensee may initially mount a gauge if permitted by the certificate of registration issued by the U.S. Nuclear Regulatory Commission or an Agreement State and under the following conditions:

    A. the gauge must be mounted in accordance with written instructions provided by the manufacturer;

    B. the gauge must be mounted in a location compatible with the "Conditions of Normal Use" and "Limitations and/or Other Considerations of Use" in the certificate of registration issued by the U.S. Nuclear Regulatory Commission or an Agreement State;

    C. the on-off mechanism (shutter) must be locked in the off position, if applicable, or the source must be otherwise fully shielded;

    D. the gauge must be received in good condition (i.e., package was not damaged); and

    E. the gauge must not require any modification to fit in the proposed location.

Mounting does not include electrical connection, activation or operation of the gauge. The source must remain fully shielded and the gauge may not be used until it is installed and made operational by a person specifically licensed by the U.S. Regulatory Commission or an Agreement State to perform such operations.

19. A. The licensee may maintain, repair, or replace device components that are not related to the radiological safety of the device containing byproduct material and that do not result in the potential for any portion of the body to come into contact with the primary beam or in increased radiation levels in accessible areas.

    B. The licensee may not maintain, repair, or replace any of the following device components: the sealed source, the source holder, source drive mechanism, on-off mechanism (shutter), shutter control, or shielding, or any other component related to the radiological safety of the device, except as provided otherwise by specific condition of this license.

20. Prior to initial use and after installation, relocation, dismantling, alignment, or any other activity involving the source or removal of the shielding, the licensee shall assure that a radiological survey is performed to determine radiation levels in accessible areas around, above, and below the gauge with the shutter open. This survey shall be performed only by persons authorized to perform such services by the U.S. Regulatory Commission or an Agreement State.

NRC FORM 374A	U.S. NUCLEAR REGULATORY COMMISSION	PAGE 6 of 6 PAGES
		License Number 08-00000-01
MATERIALS LICENSE SUPPLEMENTARY SHEET		Docket or Reference Number 030-00000
		Amendment No. XX

21. The licensee shall operate each device containing licensed material within the manufacturer's specified temperature and environmental limits such that the shielding and shutter mechanism of the source holder are not compromised.

22. The licensee shall assure that the shutter mechanism of each device is locked in the closed position during periods when a portion of an individual's body may be subject to the direct radiation beam. The licensee shall review and modify, as appropriate, its "lock-out" procedures whenever a new device is obtained to incorporate the device manufacturer's recommendations.

23. Except for maintaining labeling as required by 10 CFR Part 20, or 71, the licensee shall obtain authorization from the U.S. Nuclear Regulatory Commission before making any changes in the sealed source, device or source-device combination that would alter the description or specifications as indicated in the respective certificate of registration issued either by the Commission pursuant to 10 CFR 32.210 or by an Agreement State.

24. In addition to the possession limits in Item 8, the licensee shall further restrict the possession of licensed material to quantities below the minimum limit specified in 10 CFR 30.35(d) for establishing decommissioning financial assurance. **(Do NOT use this license condition if applicant provides evidence of financial assurance.)**

25 The licensee is authorized to transport licensed material in accordance with the provisions of 10 CFR Part 71, "Packaging and Transportation of Radioactive Material."

26. Except as specifically provided otherwise in this license, the licensee shall conduct its program in accordance with the statements, representations, and procedures contained in the documents, including any enclosures, listed below. The U.S. Nuclear Regulatory Commission's regulations shall govern unless the statements, representations, and procedures in the licensee's application and correspondence are more restrictive than the regulations.

    A.   Application dated [insert date]
    B.   Letter dated [insert date]

For the U.S. Nuclear Regulatory Commission

Date [insert license issue date]   By   Original signed by [insert reviewer's name]

MATERIALS LICENSE
SUPPLEMENTARY SHEET

License Number
08-00000-01

Docket or Reference Number
030-00000

Amendment No. XX

[insert reviewer's name]
[insert reviewer's NRC address]

Appendix R

List of Documents Considered in Development of this NUREG

List of Documents Considered in Development of this NUREG

Draft Regulatory Guides (DRGs) and Policy and Guidance Directives (P&GDs)

*FC 404-4	Guide for the Preparation of Applications for Licenses for the Use Sealed Sources in Fixed Gauging Devices	01/85
*FC 85-4	Standard Review Plan for Applications for the Use of Sealed Sources in NonPortable Gauging Devices	02/06/85
*FC 85-8	Revision 1; Licensing of Fixed Gauges and Similar Devices	06/29/88

Information Notices (INs)

IN 81-37	Unnecessary Radiation Exposures to the Public and Workers During Events Involving Thickness and Level Measuring Devices	12/15/81
IN 86-31	Unauthorized Transfer and Loss of Control of Industrial Nuclear Gauges	05/05/86
IN 88-02	Lost or Stolen Gauges	02/02/88
IN 88-90	Unauthorized Removal of Industrial Nuclear Gauges	11/22/88
IN 89-25	Revision 1, Unauthorized Transfer of Ownership or Control of Licensed Activities	12/07/94
IN 94-15	Radiation Exposures during an Event involving a Fixed Nuclear Gauge	03/02/94
IN 96-28	Suggested Guidance Related to Development and Implemetation of Corrective Action	05/01/96
IN 97-30	Control of Licensed Material during Reorganizations, Employee-Management Disagreements, and Financial Crises	06/03/97

Technical Assistance Requests (TARs)

1.	*TN Technologies, Inc. - Fixed Gauge Manufacturer request to allow Customers to Install Fixed Gauges	06/11/90
2.	U.S. Air Force - Multiple Questions on Moving Generally Licensed Fixed Gauges from one Use Location to Another	05/29/91
3.	Oceantrawl - Licensee Request to Convert Specifically Licensed Gauge to a General License	10/08/92

4.   Sharon Steel - Request for Review of the Adequacy of a Fixed Gauge   01/14/93
        Training Program

5.   Pennsylvania Power and Light Company - Request for Relief from six-   12/28/92
        Month Requirement to check on/off mechanism

6.   Ram Services, Inc. - Request for Authorization to Collect and Analyze   06/21/95
        Sealed Source Leak Tests

7.   Philip Morris, USA - Request for Clarification of Regulatory   03/20/96
        Requirements for Possession and Use of Gauging Devices Initial
        Distributed under the Provision of 10 CFR 31.5 and 32.51

Other Documents

1.   *Memorandum from W. L. Axelson to Vandy L. Miller Re: Requiring   07/09/86
        Shutter Operability Tests for Specifically Licensed Fixed Gauges

2.   *Memorandum from John Glenn to Nuclear Materials Branch Chiefs,   09/14/90
        Re: Installation of Fixed Gauges

3.   *Memorandum from John W. Hickey Re: Implementation of   08/12/97
        10 CFR 30.36(d)

(*) Marked items have been incorporated by this NUREG and are superseded.

Appendix S

Addendum: Responses to Comments on Draft NUREG - 1556, Vol. 4, Dated October 1997

Addendum: Responses to Comments on Draft NUREG - 1556, Vol. 4, Dated October 1997

Table S.1   Homestake Mining Company Comment, Dated January 12, 1998

Location	Subject	Comment
Appendix N	Dosimetry for non-routine operations	Our facility uses density gauges which are mounted on pipe lines. We have permission to perform non-routine maintenance which initially meant installation and today we are concerned with relocation. Our procedures are in place and a survey meter has always been used to ensure the safety of the person responsible for handling the gauge. We have never used whole body or extremity monitoring devices when performing these activities and feel the requirement is unnecessary to ensure the safety of people. If the gauge shutter is locked out, surveyed, and the task is monitored and documented as stated in Appendix J, there should be no reason for mandatory dosimetry. Granted, maintenance activities involving components related to the radiological safety of the gauge is a separate issue. We are only concerned with the relocation which may be necessary if a pipe must be changed and the gauge must be unbolted, removed, and then rebolted to the new pipe. We would hope that at a minimum, the NRC would accept a well documented procedure to suffice when the necessity to change a pipe is unpredictable and unplanned. If the NRC is adamant in enforcing mandatory dosimetry during relocation, we could agree with dosimetry when process/pipe changes are pre-planned. However, we urge you to consider the circumstances surrounding emergency process disruptions requiring relocation.

NRC Staff Response:

- 10 CFR 20.1502 requires each licensee to monitor occupational exposure to radiation and supply and require the use of individual monitoring devices by adults likely to receive, in 1 year, from sources external to the body, a dose in excess of 10% of the limits in 10 CFR 20.1201(a). A licensee need not monitor an individual's occupational exposure or supply and require the use of individual monitoring devices if the licensee can demonstrate that an individual is not likely to receive, in one year, a radiation dose in excess of 10% of the allowable limits.

- The 2nd paragraph in the "**Discussion**" section of chapter 8.14 (8.10.4 in final NUREG), item 10, addressing occupational dosimetry for non-routine operations has been revised as follows: "Individuals who perform non-routine operations such as installation, initial radiation survey, repair, and maintenance of components related to the radiological safety of the gauge, gauge relocation, replacement, and disposal of sealed sources, alignment, or removal of a gauge from service are more likely to exceed 10% of the limits as shown in Figure 8.4. Applicants may be required to provide dosimetry (whole body and perhaps extremity monitors) to individuals performing such services and must perform a prospective evaluation demonstrating that unmonitored individuals performing such non-routine operations are not likely to receive, in one year, a radiation dose in excess of 10% of the allowable limits as shown in Figure 8.4."

- Appendix N was revised as follows:
  - The 4th paragraph was changed to read: "Thus, applicants wishing to perform non-routine operations must use personnel with special training and follow appropriate procedures consistent with the manufacturer's instructions and recommendations that address radiation safety concerns (e.g., use of radiation survey meter, shielded container for the source, and personnel dosimetry (if required))."
  - Under the section for submission of procedures for non-routine operations, the 7th bullet was changed to read: "Confirm that individuals performing non-routine operations on gauges will wear both whole body and extremity monitoring devices or perform a prospective evaluation demonstrating that unmonitored individuals performing such non-routine operations are not likely to receive, in one year, a radiation dose in excess of 10% of the allowable limits."

Note: Unless otherwise noted, page numbers and section numbers correspond to those in Draft NUREG - 1556, Vol. 4, dated October 1997

Table S.2 J. C. Van Horn and Associates, Inc. Comments, Received on February 6, 1998

Page	Subject	Comment
xi	Abbreviations	• The Roentgen is no longer an accepted unit and should not be used in this document. This includes the abbreviations R and mR. • LDE (Lens Dose Equivalent) and DDE (Deep Dose Equivalent) need to be addressed as they are certainly exposure possibilities that may have to be considered. • CEDE (Committed Effective Dose Equivalent) and CDE (Committed Dose Equivalent) need to be addressed due to the potential for ingestion, inhalation or injection m accident conditions. Note that these are referenced on page 1-3 **Without definitions**.

NRC Staff Response:

• Fixed gauges are not typically the primary activity for the vast majority of fixed gauge licensees in the U.S. Being up-to-date on changes in health physics nomenclature would place an unreasonable burden on these licensees and would be inconsistent with the NRC's final metrication policy that requires NRC to use ". . .the system of units employed by the licensee" (*see* 61 FR 31169, June 19, 1996). NRC made a conscious decision to use the Roentgen unit in NUREG-1556, Vol. 4 to make the guide more efficient and useable for the vast majority of gauge users.
• Although Lens Dose Equivalent (LDE) and Deep Dose Equivalent (DDE) are exposure possibilities, it was decided not to address them in NUREG-1556, Vol. 4.
• Reference to Committed Effective Dose Equivalent (CEDE) and Committed Dose Equivalent (CDE) is made on page 1-3 of Draft NUREG-1556, Vol. 4. It specifically refers the reader to 10 CFR Part 20 for definitions. It states, "In this document, dose or radiation dose means absorbed dose, dose equivalent, effective dose equivalent (EDE), committed dose equivalent (CDE), committed effective dose equivalent (CEDE), or total effective dose equivalent (TEDE). These terms are defined in 10 CFR Part 20."

Page	Subject	Comment
1-3	Paragraph #4 sentence 1	Dose units mentioned do not include LDE and DDE and do include CEDE and CDE which are not previously defined

NRC Staff Response: See response to 2nd and 3rd bullets in previous comment.

Page	Subject	Comment
1-3	Paragraph #4 sentence 2	Are we sure people have access to 10CFR?

NRC Staff Response: 10 CFR Parts are readily available to license applicants and licensees. Individual copies of Parts may be obtained, by request, free of charge from the NRC and may also be "downloaded" from the NRC's Home Page on the Internet. 10 CFR Parts are also available from the Public Document Room. When an individual requests application materials from the NRC, appropriate documents including applicable 10 CFR Parts are sent to the applicant. Licensees are required by 10 CFR 19.11 to post (and thus possess) copies of Parts 19 and 20.

Page	Subject	Comment
1-3	Paragraph #4 sentence 3	Not all sealed sources only emit beta & gamma. There are also neutron and alpha sources

NRC Staff Response: The last paragraph of Chapter 1 has been revised. It now addresses neutrons and alpha particles. Please see the revised paragraph in the response to the next comment.

Page	Subject	Comment
1-3	Paragraph #4 sentence 4	Because of the use of neutron and alpha (as well as beta) sources, the assumption that 1 Roentgen = 1 rad= 1 rem is **NOT** true for many gauges due to the Quality Factors which can range up to 20.

NRC Staff Response: The last paragraph of Chapter 1 has been revised to address neutrons, alpha particles and their respective quality factors (Q). It now reads:

In this document, dose or radiation dose means absorbed dose, dose equivalent, effective dose equivalent (EDE), committed dose equivalent (CDE), committed effective dose equivalent (CEDE), or total effective dose equivalent (TEDE). These terms are defined in 10 CFR Part 20. Rem, and its SI equivalent Sievert (1 rem = 0.01 Sievert (Sv)), are used to describe units of radiation exposure or dose. This is done because 10 CFR Part 20 sets dose limits in terms of rem, not rad or roentgen (R). When the sealed sources used in gauges emit beta and gamma rays, for practical reasons, we are assuming that 1 R = 1 rad = 1 rem. Less common are sealed sources used in gauges that emit neutrons or alpha particles. For neutron and alpha emitting sealed sources, 1 rad is not equal to 1 rem. Determination of dose equivalent (rem) from absorbed dose (rad) from neutrons and alpha particles requires the use of an appropriate quality factor (Q) value. Q values are used to convert absorbed dose (rad) to dose equivalent (rem). Q values for neutrons and alpha particles are addressed in the Tables 1004(b)(1) and (2) in 10 CFR §20.1004.

Page	Subject	Comment
1-3	Paragraph #4, sentence 4	Note that quality factor is also not defined in this document.

NRC Staff Response: The last paragraph of Chapter 1 has been revised. It now states that quality factor values for neutrons and alpha particles are addressed in Tables 1004(b)(1) and (2) in 10 CFR 20.1004. In addition, the abbreviation "Q" has been added to the list of abbreviations in NUREG-1556, Vol. 4.

Page	Subject	Comment
1-3	Paragraph #4, sentence 4	The above comments on page 1-3 imply that this document's fundamental underlying assumptions may be incorrect. If this is true, then the entire document needs to be dealt with.

NRC Staff Response: See previous responses for comments on Page 1-3.

Page	Subject	Comment
3-1	Paragraph #2, bullet 4	The regulations are not specified here or prior to this point.

NRC Staff Response: NRC staff have decided to specify applicable regulations in Chapter 4 of NUREG-1556, Vol. 4. In addition, each section of the document which address the contents of the application begins by stating regulations applicable to the respective section.

Page	Subject	Comment
3-1	Paragraph #2	There is no reference to management's responsibility to ensure employees who may be exposed to radiation are properly trained in radiation hazards (10 CFR 19). As well as management responsibilities to ensure gauge users are properly trained in their use and the radiological hazards involved (10 CFR 20).

NRC Staff Response:

- See "Training Assessment" in Appendix G of the document. In draft NUREG-1556, Vol. 4 this section stated: "Management will ensure that potential RSOs and AUs are qualified to work independently with each type of gauge possessed." The final NUREG-1556, Vol. 4 states, "Management will ensure that proposed AUs are qualified to work independently with each type of gauge with which they may work. Management will ensure that proposed RSO's are qualified to work independently with and are knowledgeable of the radiation safety aspects of all types of gauges to be possessed by the applicant.

- The draft NUREG-1556, Vol. 4 had a chapter that was entitled "Training for Authorized Users and Individuals who in the Course of Employment are Likely to Receive Occupational Doses of Radiation in Excess of 1 mSv (100 mrem) in a Year." The final NUREG-1556, Vol. 4 will clarify training by dividing that chapter into two chapters: One chapter entitled, " Training for Individuals Who in the Course of Employment are Likely to Receive Occupational Doses of Radiation in Excess of 1 mSv (100 mrem) in a Year (Occupationally Exposed Workers) and Ancillary Personnel," specifically addresses training for employees who may be exposed to radiation. The other chapter entitled, "Authorized Users," specifically addresses training for gauge uses.

- The list of responsibilities (bullets) in the section of the document entitled "Management Responsibility" is not meant to provide detailed responsibilities. Rather it is meant to provide an overview of responsibilities. The responsibilities in the first bullet encompass management's responsibility to ensure that workers are properly trained. (Radiation safety, security and control of radioactive materials, and compliance with regulations.)

Page	Subject	Comment
4-1	Entire section	This section should be inserted in this document prior to any references to applicable regulations.

NRC Staff Response: NRC staff have decided to specify applicable regulations in Chapter 4 of NUREG-1556, Vol. 4.

Page	Subject	Comment
8-6	Table 8.1	This table is missing other isotopes which are in use in some gauges.

NRC Staff Response:

- The "**Discussion**" section of chapter 8.6, item 5, just prior to the table states, "The thresholds for typical radionuclides used for fixed gauge sealed sources are shown in Table 8.1." The NUREG does not seek to address all radionuclides that may be used in sealed sources for fixed gauges. As the sentence **"Discussion"** states, the table provides information for *typical* radionuclides.
- The table has been retitled: Examples of Minimum Inventory Quantity Requiring Financial Assurance.

Page	Subject	Comment
8-6	Paragraph #3	This paragraph as constituted, presents a problem for gauge holders who have been long-term users of radioisotopes. It is very possible that buildings and set-ups within process lines have been removed, renovated many years prior to the institution of this regulation. If there is no way to reconstruct the make up of the area. What provisions are allowed? What will be done for these situations that is reasonable?

NRC Staff Response: The last sentence of 10 CFR 30.35(g)(2) states, "If drawings are not available, the licensee shall substitute appropriate records of available information concerning these areas and locations." The regulation specifically stipulates *"available"* information. The regulation only requires that the licensee retain information that is available. To make this clearer, the following changes have been made:

- The 4th sentence in Paragraph #3 on page 8-6 of draft NUREG-1556, Vol. 4 has been changed to read: "If drawings are not available, licensees shall substitute appropriate records (e.g., a sketch of the room or building or a narrative description of the area) concerning the specific areas and locations. If no records exist regarding structures and equipment where gauges were used or stored, licensees shall make all reasonable efforts to create such records based upon historical information (e.g. employee recollections)."

Page	Subject	Comment
8-8	8.8 Item 7 paragraph 1 bullet 1 & 2	We have serious concerns about this provision and similar provisions throughout this document. • First, this provision ignores the Health Physics Profession and could even be seen as attacking it by removing the capability of Health Physicists to supervise radiological materials based on their standards of training. • Second, this provision improperly disqualifies personnel who are the best trained in radiation safety (HP's, Medical Physicists, and Radiation Protection Technicians). • Third, we find that this could be interpreted to disqualify even NRC HP's who might inspect these gauges. After all, if an HP isn't qualified to be an RSO, how can they inspect for radiological safety? • Fourth, this provision eliminates all other RSO Training courses. It gives a monopoly to Fixed Gauge Manufacturers in an area where they are, by far, **NOT** the most experienced personnel in dealing with Radiation Safety.

NRC Staff Response:

- The "**Criteria**" section of chapter 8.8, item 7, in the second bullet, states, in part, that an, equivalent course that meets Appendix G criteria would provide evidence of adequate training and experience for an RSO. Any individual, including health physicists, who provides evidence of training and experience that meets Appendix G criteria would qualify as an RSO for a fixed gauge licensee.
- The "**Criteria**" section of chapter 8.8, item 7, disqualifies only those individuals who do not have sufficient training and experience to serve as an RSO for a fixed gauge licensee.
- Chapter 8.8, item 7 describes training and experience for RSOs for fixed gauge licensees. NRC inspectors must complete certain training and experience to qualify as inspectors. Such training is independent of that required for qualification as an RSO for a fixed gauge licensee.
- Although the first bullet in the "**Criteria**" section of chapter 8.8, item 7, states that a fixed gauge manufacturer's course for users or for RSO would provide evidence of adequate training and experience for an RSO, it is only one of the methods an applicant may use to demonstrate evidence of adequate training and experience. The second bullet provides another method, stating, in part, that an, **equivalent** course that meets Appendix G criteria would also provide evidence of adequate training and experience for an RSO. In addition, the "**Response from Applicant**" section of chapter 8.8, item 7, states that "alternative information demonstrating that the proposed RSO and any future RSO are qualified by training and experience" will be accepted for review and evaluated using the criteria stated in the chapter.

Page	Subject	Comment
8-10	8.9 Item 8 Paragraph #1 bullets 1 & 2	• Much research indicates that there are no such courses. • Also there are no requirements for the personnel operating the gauges to be able to take surveys and protect themselves from the radiation hazard. This **MAY** contradict 10 CFR 20 requirements.

NRC Staff Response:

• Many fixed gauge manufacturers and distributors provide training for users.
• As stated in the "**Discussion**" section of chapter 8.13, item 10, "Usually it is not necessary for fixed gauge licensees to possess a survey meter." The chapter also discusses specific circumstances when surveys are required. If a survey pursuant to 10 CFR 20.1501 is required, the licensee must comply. The survey may not be adequate if the individual performing the survey is not properly trained on the use of survey meters and is thus unable to operate it properly. It is not within the scope of this document to address specific training on the use of survey instruments. However, Appendix G, "Criteria for Acceptable Training Courses for Authorized Users and Radiation Safety Officers" has been modified to add training on:
  – Use of survey meters and personal dosimetry, when required

8-11	Paragraph #1	This paragraph should be more closely compared to 10 CFR 19 requirements. "To minimize the potential" sets up the licensee for both regulatory and legal problems. e.g. "Do I or don't I train on the Rad Symbol?" "What about posting and area control?" etc.

NRC Staff Response:

• There is no specific requirement to provide training to individuals who in the course of employment are **not** likely to receive in a year an occupational dose in excess of 100 mrem (this includes the majority of gauge users). Posting requirements are addressed in 10 CFR 19.11 and 20.1902. Control of exposure from external sources in restricted areas is addressed by 10 CFR 20.1601. Storage and control of licensed material is addressed by 10 CFR 20.1801 and 20.1802. A licensee may determine that it is necessary to provide some training to individuals, even if they are not likely to receive a dose in excess of 100 mrem, to meet the requirements of 10 CFR 19.11, 20.1902, 20.1601, 20.1801 and 20.1802.
• Some modification to the "Discussion" section of 8.9 Item 8: "Training for Individuals Who in the Course of Employment Are Likely to Receive Occupational Doses of Radiation in Excess of 1 mSv (100 mrem) in a Year (Occupationally Exposed Workers) and Ancillary Personnel," was made to clarify this issue.

Page	Subject	Comment
8-13	8.13 Item 10 Paragraph 2 bullets 1.2 & 3	• Use of mR vs. mrem (mR & R undefined in 10CFR20) • An instrument capable of measuring gamma radiation isn't going to be much use with neutron or alpha radiation, and some gamma-sensitive instruments won't be useful for beta • This requires source checks, how about the other operational checks? - physical condition, battery check etc.?

NRC Staff Response:

• Bullet 1 - The vast majority of exposure rate survey instruments measure in mR and R units. The bullet does provides the alternative unit (C/kg). The last paragraph in chapter 1 states that "10 CFR Part 20 sets dose limits in terms of rem, not rad or roentgen (R)." This implies that the R is undefined in 10 CFR Part 20. The last paragraph in chapter 1 has also been modified to address mR vs. mrad vs. mrem.
• Bullet 2 - This bullet has been modified to address this comment. It now reads "Is capable of measuring the radiation being emitted from the gauge's sealed source."
• Bullet 3 - A functionality check should indicate if the battery is not functioning. Other operational checks are addressed during calibration - see bullet 4.

Page	Subject	Comment
8-14 &15	All	No training on how to use these instruments is required.

NRC Staff Response: Appendix G, "Criteria for Acceptable Training Courses for Authorized Users and Radiation Safety Officers" has been modified to add training on:

• Use of survey meters and personal dosimetry, when required

Page	Subject	Comment
8-18	Figure 8.4	• Note LDE (although not named as such) is included here, but not referenced elsewhere. • Missing Internal dose (CEDE & CDE) • Missing DDE

NRC Staff Response: See previous NRC Staff responses on these dose equivalent comments.

Page	Subject	Comment
8-18	Paragraphs #1, 2, 3	• What is the basis for the assumption that dosimetry will not be required? • If dosimetry is required, there is no requirement for training in it's use.

NRC Staff Response:

• In the "**Discussion**" section of chapter 8.14 (8.10.4 in final NUREG), item 10, it states, "Under conditions of routine use, the typical fixed gauge user does not require a personnel monitoring device (dosimetry)." The basis for the statement is NRC inspection experience which indicates that under conditions of routine use, the typical fixed gauge user is not likely to receive, in one year, a radiation dose in excess of 10% of the allowable limits specified in 10 CFR 20.1201(a). 10 CFR 20.1502 requires each licensee to monitor occupational exposure to radiation and supply and require the use of individual monitoring devices by adults likely to receive, in 1 year, from sources external to the body, a dose in excess of 10% of the limits in 10 CFR 20.1201(a).

• It is not within the scope of this document to address specific training on the use of personal monitoring devices. However, Appendix G, "Criteria for Acceptable Training Courses for Authorized Users and Radiation Safety Officers" has been modified to add training on:
  – Use of survey meters and personal dosimetry, when required

Page	Subject	Comment
8-22	Routine Procedures bullet 8	• How does an individual become "authorized" by the NRC to do non-routine repair & maintenance on a gauge? • What are the qualifications, training and experience requirements. • Is professional Radiation Protection Training and/or experience necessary?

NRC Staff Response:

• Chapter 8.18 (8.10.8 in final NUREG), Radiation Safety Program - Maintenance, in the "**Criteria**" sections, states, "Information to support request for specific authorization to perform non-routine maintenance or repair is addressed in Appendix N." This information will be reviewed on a case-by-case basis; if approved, the license will contain a condition authorizing the licensee to perform non-routine operations.

• Appendix N, asks the applicant to, "Identify who will perform non-routine operations and their training and experience. Acceptable training would include manufacturer's or distributor's courses for non-routine operations or equivalent." Submissions will be reviewed by NRC license reviewers on a case-by-case basis.

• Appendix N, states, "Acceptable training would include manufacturer's courses for non-routine operations or equivalent." Submissions will be reviewed by NRC license reviewers on a case-by-case basis.

Page	Subject	Comment
8-22	Emergency Procedures bullet	Need to require control of potentially contaminated personnel and objects.

NRC Staff Response: A detailed consideration of every possible emergency situation involving a fixed gauge that could occur is beyond the intended scope of chapter 8.16 (8.10.6 in final NUREG), item 10. One of the bullets under emergency procedures, in the "**Criteria**" section states, "Take additional steps, dependent on the specific situations."

Page	Subject	Comment
8-24	Figure 8.6	Why do you talk about trained radiological personnel here and ignore their existence and qualifications elsewhere?

NRC Staff Response: The last illustration (#8) in Figure 8.6, has a caption which states, "Trained Radiological Professional Handle Gauge Recovery and Clean-up." Use of fixed gauges is not typically the primary activity for the vast majority of fixed gauge licensees. Fixed gauge licensees be capable of responding to an emergency situation involving a fixed gauge. This caption is meant to imply that a typical fixed gauge user or RSO may not have the resources or the training and experience to handle the aftermath of a incident where a gauge is damaged, the sealed source may be exposed, and contamination is a possibility. More experienced individuals may need to be consulted to ensure that recovery and clean-up is performed safely.

Page	Subject	Comment
8-29	Figure 8.9	• Why does this picture/diagram suggest that NRC authorization is necessary to do non-routine maintenance on the gauge? • We guess that it's intent is to indicate that the person doing the maintenance is authorized by the NRC to do so, but again, how does one obtain this authorization?

NRC Staff Response:

- The non-routine maintenance portion of the "**Response from Applicant**" section of chapter 8.18 (8.10.8 in final NUREG), item 10, states, in part, that the applicant must commit that either the gauge manufacturer or other person authorized by the NRC or an Agreement State will perform non-routine maintenance, or the applicant must submit the information in Appendix N of NUREG-1556, Vol. 4 for NRC review.
- The applicant may obtain NRC authorization to perform non-routine maintenance by submitting the information listed in Appendix N for NRC review and receiving an amendment authorizing the licensee to perform non-routine maintenance.

Page	Subject	Comment
8-28 & 29	Item in box on 8-28 and 8 bullets discussion on 8-29	These two are repetitious and one should be eliminated with any "missing elements" incorporated into the one which is retained.

NRC Staff Response: NRC Staff believe that the information in the box in the "**Criteria**" section of chapter 8.18 (8.10.8 in final NUREG), item 10 and the bullets in the "**Discussion**" section of the same chapter are both necessary to the document's content and should be retained.

Page	Subject	Comment
8-33	Figure 8-11	Need to deal with contamination control measures as referenced before.

NRC Staff Response:

- A detailed consideration of every possible emergency situation involving a fixed gauge that could occur is beyond the intended scope of chapter 8.16 (8.10.6 in final NUREG), item 10. The 3rd bullet under emergency procedures, in the "**Criteria**" section states, "Contact responsible individuals…" The last bullet states, "Take addition steps, dependent on the specific situations."
- Appendix L, under emergency procedures, does address contamination control measures.

Page	Subject	Comment
8-39	8.21 Item 10	Needs to address contamination control in emergency or accident situations.

NRC Staff Response:

- A detailed consideration of every possible emergency situation involving a fixed gauge that could occur is beyond the intended scope of chapter 8.16 (8.10.6 in final NUREG), item 10. The 3rd bullet under emergency procedures, in the "**Criteria**" section states, "Contact responsible individuals…" The last bullet states, "Take addition steps, dependent on the specific situations."
- Appendix L, under emergency procedures, does address contamination control measures.

Page	Subject	Comment
B-3	Item 8	An item entitled "Training for Potentially Exposed Personnel" needs to be added. This should include the items previously discussed in these comments and in the document.

NRC Staff Response: The draft NUREG-1556, Vol. 4 had a chapter that was entitled "Training for Authorized Users and Individuals who in the Course of Employment are Likely to Receive Occupational Doses of Radiation in Excess of 1 mSv (100 mrem) in a Year." The final NUREG-1556, Vol. 4 will clarify training by dividing that chapter into two chapters: One chapter entitled, "Training for Individuals Who in the Course of Employment are Likely to Receive Occupational Doses of Radiation in Excess of 1 mSv (100 mrem) in a Year (Occupationally Exposed Workers) and Ancillary Personnel," specifically addresses training for employees who may be exposed to radiation. The other chapter entitled, "Authorized Users," specifically addresses training for gauge uses. Appendix B in the final NUREG-1556, Vol. 4 has been appropriately modified to reflect the changes in the text of these two new chapters.

Page	Subject	Comment
B-4	Item 10 Survey Instruments	Need to add an item entitled "Training For Personnel Who Use Survey Instruments". This training should include the standard instrument use training such as enter the area with device on lowest setting, source check the instrument, operationally check the instrument (battery check etc.), proper survey techniques, etc.

NRC Staff Response: It is not within the scope of this document to address specific training on the use of survey instruments. However, Appendix G, "Criteria for Acceptable Training Courses for Authorized Users and Radiation Safety Officers" has been modified to add training on:

• Use of survey meters and personal dosimetry, when required

Page	Subject	Comment
B-4	Item 10 Dosimetry	• Need to add an item about training on proper use of dosimetry. This should include care, positioning, precautions etc. • Note that the numbering system could be confusing.

NRC Staff Response: It is not within the scope of this document to address specific training on the use of personal monitoring devices. However, Appendix G, "Criteria for Acceptable Training Courses for Authorized Users and Radiation Safety Officers" has been modified to add training on:

• Use of survey meters and personal dosimetry, when required
• The numbering system is meant to correspond to information requested in NRC Form 313 in Appendix A of NUREG-1556, Vol. 4.

Page	Subject	Comment
D-2 & D-3	Materials Possessed	Should provide for more isotopes.

NRC Staff Response: There is a row in items 5 and 6 that states "Other isotope (specify):" Isotopes not specifically listed in items 5 and 6 may be added in this row.

Page	Subject	Comment
D-4	RSO Qualifications	What about Health Physicists? Also what about retraining frequency & content requirements? etc.?

NRC Staff Response:

- Health Physicists are not precluded from providing their training and experience in support of a request to name an RSO. Item 7 of the reviewer checklist, Appendix D, reflects information covered in chapter 8.8, item 7 which describes training and experience for RSOs for fixed gauge licensees. There are several methods that provide evidence of adequate training and experience for an RSO. Completion of a fixed gauge manufacturer's course for users or for RSOs or an **equivalent** course that meets Appendix G criteria would each provide evidence of adequate training and experience for an RSO. In addition, the checklist allows an applicant to check the "Other- Yes" column and submit alternative information demonstrating that the proposed RSO and any future RSO are qualified by training and experience. Alternative information will be accepted for review and evaluated using the criteria stated in chapter 8.8, item 7.
- 10 CFR 19.12(a)(1) states that, "All individuals who in the course of employment are likely to receive in a year an occupational dose in excess of 100 mrem (1 mSv) shall be — **kept informed** of the storage, transfer, or use of radiation and/or radioactive material. The phrasing "kept informed" may necessitate periodic retraining for this group of individuals. Other retraining may be necessary in order for licensees to ensure that their workers have the skills to meet the requirements of NRC regulations (e.g., security and control of materials, surveys, posting).

Page	Subject	Comment
D-5	Individual responsible for training	• These qualifications are inaccurate and inadequate. They don't reflect the capabilities of Health Physicists and other Radiation Protection Trained personnel. • Note that the operation of such a gauge is **NOT DIFFICULT**, thus the Radiation Safety aspects of this issue are paramount. • Note that a "Physical Science or Engineering Degree" could include Astronomy, Geology, Paleontology. • What about Health Science Degrees such as X-Ray Technician? They have far more radiation safety training than a geologist! • What happened to the "Or Equivalent" clause?

NRC Staff Response:

- The NRC Staff believe the qualifications for course instructors are adequate. Alternative qualifications may be submitted. NRC license reviewers will review alternative qualifications on a case-by-case basis.
- The NRC Staff's concern is with radiation safety of the gauge. The NRC Staff note that the operation of many gauges is not difficult.
- A physical science or engineering degree could include astronomy, geology or paleontology. However, "Course Instructor Qualifications" for individual's with a Bachelor's degree in a physical or life science also suggests successful completion of both a fixed gauge manufacturer's course for users, an 8 hour radiation safety course and 8 hours of hands-on experience with fixed gauges.
- If individuals with health sciences degrees such as X-ray technicians meet one of the training options stated in Item 7 "Individual(s) Responsible for Radiation Safety Program and Their Training and Experience - Radiation Safety Officer" of Appendix D, then they may qualify to be an RSO.
- In the "Purpose of the Report" Chapter there is a bullet which instructs:
  - Response from Applicant — provides suggested response(s), offers the option of an alternative reply, or indicates that no response is needed on that topic during the licensing process.

The last two columns in the checklist state "Other - Yes or No." If the licensee checks "Yes" in the "Other" column, they may attach an alternative reply to the application. NRC license reviewers will review alternative replies on a case-by-case basis.

Page	Subject	Comment
D-6 D-7	RSO Training for Authorized Users	Again, what about Health Physics Qualifications? Does the item for "Annual audit of radiation safety program" and it's sub-bullets really make sense here? This is an RSO Responsibility, not an "Authorized User's"

NRC Staff Response:

- Health Physicists qualifications will be reviewed against the criteria stated in NUREG-1556 Vol. 4 or the applicant may elect to submit the qualifications of a health physicist as an alternative reply.
- The NRC Staff believes that authorized users should be aware of the requirement for an annual audit of the program.

Page	Subject	Comment
D-8	Training for Authorized Users	• Bullet 4 - how about "Each gauge the Authorized User" will use? some licensees possess several gauges that an individual will never use due to their job description & location. • What about training on the use of survey meters? • What about retraining frequency and content requirements? • Course Instructor qualifications are faulty as previously described. These **HAVE** to be addressed.

NRC Staff Response:

- Bullet 4 has been changed to read "Practical Explanation of the Theory and Operation for Each Type of Gauge that may be used by the Authorized User"
- It is not within the scope of this document to address specific training on the use of survey instruments. However, Appendix G, "Criteria for Acceptable Training Courses for Authorized Users and Radiation Safety Officers" has been modified to add training on:
  – Use of survey meters and personal dosimetry, when required

- 10 CFR 19.12(a)(1) states that, "All individuals who in the course of employment are likely to receive in a year an occupational dose in excess of 100 mrem (1 mSv) shall be — **kept informed** of the storage, transfer, or use of radiation and/or radioactive material. The phrasing "kept informed" may necessitate periodic retraining for this group of individuals. Other retraining may be necessary in order for licensees to ensure that their workers have the skills to meet the requirements of NRC regulations (e.g., security and control of materials, surveys, posting).
- The NRC Staff believe the qualifications for course instructors are adequate. Alternative qualifications may be submitted. NRC license reviewers will review alternative qualifications on a case-by-case basis.

Page	Subject	Comment
D-10	Instruments Optional Response	• Instruction is unclear in it's intent. • Incorrect Units mR not mrem. • Why 50 μC/kg? not used. • Measuring Gamma radiation - What happened to alpha, beta and neutron?

NRC Staff Response:

- The "criteria" section of chapter 8.13 item 10: "Radiation safety Program - Instruments" provides a description of intent with regard to instruments.
- See previous response regarding use of the Roentgen.
- NRC's final metrication policy that requires NRC to use "…the system of units employed by the licensee" (*see* 61 FR 31169, June 19, 1996). NRC made a conscious decision to use the Roentgen unit in NUREG-1556, Vol. 4 to make the guide more efficient and useable for the vast majority of gauge users.
  - 50 μC/kg per hour is stated in the first bullet under optional response in the instrument section of Appendix D.
  - See also the "Criteria" section of chapter 8.13 item 10: "Radiation Safety Program - Instruments" which refers to 50 μC/kg.

- The second bullet in the criteria" section of chapter 8.13 item 10: "Radiation safety Program - Instruments" has been changed to read, "Is capable of measuring the radiation being emitted from the gauge's sealed source."

Page	Subject	Comment
D-13	Operating & Emergency Procedures	Why is there no requirement for surveying the gauge/work area? This is a simple, basic safety precaution. Every RG Technician at a DOE Facility or a Nuclear Power Plant would wince at the lack of this & issue a "stop work order."

NRC Staff Response:

- 10 CFR 20.1501 requires each licensee to make or cause to be made, surveys that may be necessary for the licensee to comply with the regulations in this part; and are reasonable under the circumstances to evaluate the extent of radiation levels; and concentrations or quantities of radioactive material; and the potential radiological hazards that could be present. Some of the bullets listed in Appendix D - Operating and Emergency procedures, depending upon circumstances, may require that a survey of the gauge/work area be made.
- Chapter 8.13 item 10: "Radiation safety Program - Instruments" discusses surveys and instruments.
- Appendix G, "Criteria for Acceptable Training Courses for Authorized Users and Radiation Safety Officers" has been modified to add training on:
    - Use of survey meters and personal dosimetry, when required

Page	Subject	Comment
D-15	Emergency procedures	What about Contamination Control and control of contaminated personnel?

NRC Staff Response: A detailed consideration of every possible emergency situation involving a fixed gauge that could occur is beyond the intended scope of the applicant's response. The 3rd bullet under emergency procedures, states, "Contact responsible individuals..." The last bullet states, "Take additional steps, dependent on the specific situations." Appendix L, under emergency procedures, does address contamination control measures.

Page	Subject	Comment
D-17	Routine Safety Program Maintenance	• Potential Problem: manufacturer's recommendations may NOT include good radiation safety precautions. • Is each "Optional Response" a completely separate Option?

NRC Staff Response:

- Licensee's must comply with NRC regulations in addition to the manufacturer's recommendations.
- The applicant may supply an optional response for any of the items requested in the application.

Page	Subject	Comment
F-1	Bullet 4 - Training	Should have the RSO "supervise" the proper training.

NRC Staff Response: There is no requirement for the RSO to supervise training.

Page	Subject	Comment
G-1	Radiation Safety	Include proper survey techniques and response to unusual readings.

NRC Staff Response: It is not within the scope of this document to address specific training on the use of survey instruments. However, Appendix G, "Criteria for Acceptable Training Courses for Authorized Users and Radiation Safety Officers" has been modified to add training on:

- Use of survey meters and personal dosimetry, when required

Page	Subject	Comment
G-1	Theory of the Gauge & lock out procedures	Is this the "manufacturer's course (which doesn't exist)? Is normal lock-out, tag-out training sufficient?

NRC Staff Response: "Normal lock-out, tag-out training" is sufficient if it is equivalent to the manufacturer's training or equivalent. If the manufacturer or distributor does not provide such training the licensee should submit alternative training. NRC license reviewers will determine on a case-by-case basis if training is adequate.

Page	Subject	Comment
G-2	OUT	• Missing requirement for surveys for radiological safety. • Supervision requirement should allow an HP or the RSO to do this. • How do you qualify an "authorized user" if no one has had the training?

NRC Staff Response:

- As stated in the "**Discussion**" section of chapter 8.13, item 10, "Usually it is not necessary for fixed gauge licensees to possess a survey meter." This chapter also discusses specific circumstances when surveys are required. If a survey pursuant to 10 CFR 20.1501 is required, the licensee must comply.
- The training and experience criteria for RSOs qualify them as authorized users. Thus on-the-job training could also be done under the supervision of an RSO. Wording of the sentence will be changed to add "or RSO." If the health physicist's training and experience meet those for an authorized user, the health physicist may supervise on-the-job training.
- The proposed authorized user should be able to obtain training from the manufacturer or distributor of the fixed gauge. If this is not possible, the applicant should submit an alternative response for training. NRC license reviewers will review alternative responses on a case-by-case basis.

Page	Subject	Comment
G-2	Training Assessment	• Each type of gauge possessed is not practical for some users. Use "Each type of gauge an Authorized User will work with." • Qualifications need some form of formal documentation.

NRC Staff Response:

- The proposed wording in Appendix G, under "Training Assessment," for the final NUREG - 1556, Vol. 4 states, "Management will ensure that **proposed** AUs **and RSOs** are qualified to work independently with each type of gauge **with which they may work. Management will ensure that proposed RSO's are qualified to work independently with, and are knowledgeable of the radiation safety aspects of, all types of gauges to be possessed by the applicant.**

- Course instructor qualifications need not be submitted with the application. Evidence of training (e.g. documentation) may be reviewed at time of inspection.

Page	Subject	Comment
G-2	Instructor Qualifications	• Inconsistent with previous requirements • No specification of the hours of hands on experience. • Why is this "hours of hands-on" a requirement? These gauges are **NOT** "rocket science". The concern here is for radiological safety. An 8 hour course in the subject is woefully inadequate. • Bachelor's degree requirements have the same problem as before, I wouldn't want to have my radiological safety "guaranteed" by a paleontologist. • Is a Health Physics Degree or certification or "X" years (I suggest 5 years) of Health Physics Experience equivalent? • The 40 hr radiation safety course here is NOT??? the manufacturer's course? - Inconsistent!! (But it makes more sense.)

NRC Staff Response:

- There are no specific requirements for course instructor qualifications. The guidance in draft NUREG-1556, Vol. 4 combined and updated previous guidance. Thus, course instructor qualifications in NUREG-1556, Vol. 4 may be inconsistent with some previous guidance.
- Specification of eight hours of hands-on experience with fixed gauges was inadvertently omitted from the draft document.
- Alternative training and experience for course instructors may be submitted for NRC review. Appendix G has been revised to clearly state that alternative qualifications may be submitted for NRC review as follows: "The applicant may submit a description of alternative training and experience for the course instructor."
- Course instructors should have radiation safety training in addition to that provided by the manufacturer. Course instructors who do not possess an appropriate undergraduate degree may require more radiation safety training as they may not have the training background to assimilate radiation safety principles as quickly as an individual with a bachelor's degree in a physical or life science or engineering. As stated above, the applicant may submit a description of alternative training and experience for the course instructor.

Page	Subject	Comment
H-2	Training & Instructions	Lots of questions in this section that no guidance is provided for prior to this e.g.: a, b, d, h, i.

NRC Staff Response: Appendix H is the Suggested Fixed Gauge Audit Checklist. It is not the intent of the document to specifically address each NRC regulation with regard to fixed gauge licensees. Appendix H attempts to provide a fairly comprehensive list of areas that the licensee should address during an audit. Appendix H may address requirements that are not addressed elsewhere in the document.

- a is addressed in 8.9, item 8.
- b is addressed in 8.8, item 7.2.
- d is addressed in Appendix N.
- h is addressed in 8.8 item 7.2 and 8.19 (8.10.9 in final NUREG) item 10.
- i is addressed in 8.19 (8.10.9 in final NUREG), item 10, 8.20 (8.10.10 in final NUREG), item 10 and in Appendix O by reference to 49 CFR.

Page	Subject	Comment
H-2 & 3	Personnel Radiation Protection	• ALARA is required in 10CFR20 why is the emphasis not existent through the rest of the document (e.g., Survey requirements etc.)? • e. Nowhere prior to this is any "warning" of the DPW concerns given.

NRC Staff Response:

• Appendix H is the Suggested Fixed Gauge Audit Checklist. It is not the intent of the document to specifically address each NRC regulation with regard to fixed gauge licensees. Appendix H attempts to provide a fairly comprehensive list of areas that the licensee should address during an audit. Appendix H may address requirements that are not specifically addressed elsewhere in the document.
  – 10 CFR 20.1101(b) states the ALARA requirement and applies to all licensees. It requires that licensees use to the extent practicable, procedures and engineering controls based upon sound radiation protection principles to achieve occupational doses and doses to members of the public that are as low as reasonably achievable. Thus, ALARA applies to every aspect of the licensee's entire Radiation Protection Program. ALARA is addressed in the audit checklist for overall evaluation of the Radiation Protection Program. ALARA is addressed throughout the document. ALARA is defined in abbreviations and is discussed in many chapters of the document including chapters 8.11 Item 10: Audit Program; 8.16 (8.10.6 in final NUREG) Item 10: Operating and Emergency Procedures; 8.18 (8.10.8 in final NUREG) Item 10: Maintenance; and 8.20 (8.10.10 in final NUREG) Item 10: Fixed Gauges Used at Temporary Job Sites.
  – Chapter 8.13, item 10, discusses specific circumstances when surveys are required. If a survey pursuant to 10 CFR 20.1501 is required, the licensee must comply.

• 10 CFR 20.1208 addresses declared pregnant woman (DPW) and applies to all licensees. If a woman declares pregnancy, the licensee would need to ensure that all of the requirements of 20.1208 are met. This regulation is addressed in the audit checklist to advise licensees to the possibility of a DPW in their program.

| H-5 | Posting & Labeling | How about "is the Radiological Hazard properly posted"? |

NRC Staff Response: The requirements for posting radiological hazards are specified in 10 CFR 20.1902 and 20.1904. These regulations are stated in parenthesis following item 14-c on page H-5 of draft NUREG-1556, Vol. 4.

| I-1 | OJT | How is the OJT documented? |

NRC Staff Response:
A specific method of documenting on-the-job training (OJT) is not stipulated. OJT may be documented by the applicant in numerous ways and will be verified at time of inspection.

Page	Subject	Comment
I-2	Survey meter	• Must be able to detect the appropriate type of radiation emitted by the source. • Calibration must be made for this type of radiation at the correct energy.

NRC Staff Response:

The third and fourth bullet of section 1 of Appendix I have been changed to read:

• Contain a radionuclide which emits radiation of identical or similar type and energy as the sealed sources that the instrument will measure
• Be strong enough to emit a radiation field that is representative of the field being emitted by the gauge. For calibration of instruments intended to measure gamma radiation, the exposure rate should be at least 30 mR/hour (7.7 microcoulomb/kilogram per hour) at 100 cm [e.g., 3.1 gigabecquerels (85 millicuries) of Cs-137 or 780 megabecquerels (21 millicuries) of Co-60].

J-1	Example	Should scatter be considered?

NRC Staff Response: Consideration of scattered radiation is beyond the intended scope of the example provided in Appendix J.

L-1	Operating Procedures	Survey of area? It should be required periodically.

NRC Staff Response: Chapter 8.13, item 10, discusses specific circumstances when surveys are required. If a survey pursuant to 10 CFR 20.1501 is required, the licensee must comply.

L-2	Emergency Procedures	How about control of contaminated objects and personnel?

NRC Staff Response: Detailed consideration of control of contaminated objects and personnel is beyond the intended scope of the emergency procedures provided in Appendix L. However, some consideration is given to contamination control in Appendix L in the 2nd and 3rd bullets under Emergency Procedures. See also response to comments regarding figure 8.6 on page 8-24.

N-2	Identify who will perform non-routine maintenance	What is equivalent training? Especially if the manufacturer's courses don't exist?

NRC Staff Response: The applicant is instructed that acceptable training includes the manufacturer's courses on non-routine operations or equivalent training. NRC license reviewers will determine on a case-by-case basis if training on non-routine operations other than that provided by a manufacturer is adequate.